Telecommunications

Veröffentlichungen des / Publications of the

Münchner Kreis
Übernationale Vereinigung für Kommunikationsforschung,
Supranational Association for Communications Research

Band / Volume 5

Neue Formen der Datenkommunikation

New Forms of Data Communication

Vorträge des am 1./2. Juli 1980
in München abgehaltenen Symposiums

Proceedings of a Symposium
Held in Munich July 1/2, 1980

Herausgeber/Editor: G. Seegmüller

Springer-Verlag
Berlin Heidelberg New York 1981

Münchner Kreis
Übernationale Vereinigung für Kommunikationsforschung
Supranational Association for Communications Research
Ludwigstraße 8, D-8000 München 22, Telefon: (089) 28 49 09

Wissenschaftliche Betreuung des Kongresses:

Prof. Dr. Gerhard Seegmüller
Institut für Informatik der Universität München und
Leibniz-Rechenzentrum der Bayerischen Akademie
der Wissenschaften, München

Wissenschaftliche Redaktion:

Dr. Heinz-Gerd Hegering
Leibniz-Rechenzentrum der Bayerischen Akademie
der Wissenschaften, München

CIP-Kurztitelaufnahme der Deutschen Bibliothek
Neue Formen der Datenkommunikation :
Vorträge d. am 1./2. Juli 1980 in München abgehaltenen
Symposiums = New forms of data communication /
[Münchner Kreis, Übernationale Vereinigung für
Kommunikatipnsforschung]. Hrsg.: G. Seegmüller. –
Berlin ; Heidelberg ; New York : Springer, 1981.
(Telecommunications ; Bd. 5)

ISBN-13:978-3-540-10736-1   e-ISBN-13:978-3-642-81638-3
DOI: 10.1007/978-3-642-81638-3

NE: Seegmüller, Gerhard [Hrsg.]; Münchner Kreis;
PT; GT

Vorwort

Unter dem führenden Einfluß insbesondere von nordamerikanischen
Netzträgern und Herstellern gab es in den letzten Jahren eine
recht lebhafte Entwicklung auf den Gebieten der Daten- und Telekommunikation. Davon sind sowohl die Trägertechnologien selbst,
als auch die in den Netzen realisierten Dienste erfaßt worden.
Das Schlagwort von den Multifunktionsnetzen ist in aller Munde.
Im Münchner Kreis ist diese wichtige Entwicklung aufmerksam
verfolgt worden.

Was die Möglichkeiten der technischen Entwicklung anlangt, liegt
in der Tat eine einmalig günstige Situation vor, da in weitgehend voneinander unabhängigen Forschungs- und Entwicklungsgebie-
ten Ergebnisse erzielt wurden, die in kombinierter Anwendung zu
einer neuen Qualität im Bereich der Datenkommunikation führen
können. Datenkommunikation ist aber ein einheitliches technisches
Fundament, auf welchem ein breites Spektrum von Telekommunikations-
formen aufgebaut werden kann. Die Forschungs- und Entwicklungsge-
biete, die zu dieser Entwicklung beitragen, sollen kurz genannt
werden.

Die oft besprochenen und in ihrer vollen Auswirkung sicherlich
noch nicht verstandenen Fortschritte in der Halbleitertechnologie
sind auch hier mitentscheidend. Sie ermöglichen den massiven und
dennoch wirtschaftlichen Einsatz aktiver digitaler Komponenten
in Netzen. Darin ist auch die Möglichkeit zu funktionellem Reich-
tum in solchen Einrichtungen begründet.

Im Bereich der Übertragungstechnologien selbst zeichnet sich die
Möglichkeit zu wirtschaftlicher Übertragung mit wesentlich höhe-
ren Geschwindigkeiten ab. Darüber hinaus erscheint eine verein-
heitlichte digitale Behandlung von Akustik-, Video- und herkömm-
licher Digital-Information als sehr aussichtsreich. Das Element

der massiven Rechnerunterstützung verspricht neue Dimensionen
der flexiblen Netzsteuerung, der Netzverwaltung, der Fehlerbe-
handlung, des Datenschutzes und eines bislang noch nicht abzu-
sehenden Funktionsreichtums an einer sehr hohen Netzschnitt-
stelle. Das nicht nur zu reinen Übertragungszwecken speicher-
fähige Netz wird sicherlich zu einem interessanten Gegenstand
der Diskussion werden.

Die Konvergenz der Bemühungen unter Verwendung der Resultate
aus den genannten Disziplinen in Richtung auf ein gemeinsames
Netzkonzept wird erleichtert durch den fortschreitenden internationalen Normungsprozeß insbesondere in der Paketvermittlung.
Jedoch harren gerade hier noch einige Probleme einer gemeinsa-
men Lösung.

Der schon genannte, umfangreiche Rechnereinsatz zieht die Notwendigkeit der Erstellung vieler Betriebs- und Dienstprogramme
nach sich, die bei extremen Korrektheits- und Flexibilitätsfor-
derungen auch noch über ein sehr gutes Fehlerverhalten verfügen
müssen. Es gibt heute Software-Firmen, die über die dafür notwendige technische Kompetenz verfügen. Die langjährigen Bemühungen um eine grundlegende,breite akademische Ausbildung von
Informatikern zeigen hier erste Erfolge.

Der vorliegende Band enthält die Vorträge, die auf dem Symposium
des Münchner Kreises mit dem Titel "Neue Formen der Datenkommunikation in den USA" am 1. und 2. Juli 1980 in München gehalten
wurden. Angesichts der Beiträge auch aus Kanada und Europa ist
der Titel, der sehr früh während der 1 1/2jährigen Vorbereitun-
gen festgelegt wurde, zu eng. Aus dem Wunsche nach einer gewissen
Stimulation der Teilnehmer wurde er jedoch beibehalten.

Die Beiträge sind teils in Englisch, teils in Deutsch abgefaßt.
Eine Kurzfassung in der jeweils anderen Sprache ist hinzugefügt
worden. Damit folgen wir einem guten Brauch innerhalb dieser Ver-
öffentlichungsserie des Münchner Kreises.

München, im Dezember 1980

                      G. Seegmüller
                     Leiter des Symposiums

Foreword

Recently, there have been quite vigorous activities in the areas of data and telecommunications. This has been happening under the leading influence particularly of North American carriers and manufacturers. Both transmission media technology and network services were affected substantially. The buzzword of value-added networks has become almost commonplace. The Münchner Kreis has been closely watching this important development.

As to the technical possibilities there is a unique and very promising situation, indeed. By their combination new results in several independent fields might lead to a new level of quality in data communications. This very area, however, may serve as a uniform technical base upon which a broad spectrum of forms of telecommunication can be built. Let me briefly mention some of the technological fields which are contributing to this end.

There is the often mentioned and still not fully understood dramatic progress in semiconductor technology. This progress is of key importance to massive and yet economic usage of active digital components in networks. That usage, in turn, is the base for functional richness of such systems.

The economic possibility for significantly higher transmission speeds is one of the results of research in transmission media technology. There are also good prospects for a representational unification of acoustic, video, and classical digital data on a uniform digital basis. Massive computerization of networking will lead to new dimensions in flexible transmission control, network administration, fault handling, data protection, and functionally very rich and high network interfaces. The concept of a network which is capable of storing large amounts of

information beyond the purposes of mere transmission will be
a topic of considerable discussion and concern.

The ongoing process of international standardization, e.g.
in packet switching, is of quite some help for easier converg-
ence of the above tendencies towards a common and uniform networking concept. And yet it is this very area where some
difficult problems have still to be solved.

Computerization of networks implies a large amount of highest
quality operating and service software which has to be
constructed and built. It is indispensable that this software
has to have excellent fault tolerance properties apart from
the usual criteria of correctness and functional flexibility.
Fortunately, we have software companies today which can
successfully meet this technical challenge. The long standing
efforts in many computer science departments towards a broad
academic education of software engineers are beginning to show
some bearing.

This volume contains the lectures given at the Münchner Kreis
Symposium on "New Forms of Data Communication in the United
States" held in Munich on July 1 and 2, 1980. In view of
contributions also from Canada and Europe this title is too
narrow. It was chosen very early during the 1 1/2 years of
preparation and it was then kept also as a kind of stimulation
for potential participants.

Some of the lectures are in English, some in German. We have
always added a summary in the other language. This is in
accordance with the habit adopted in previous volumes of this
series of publications of the Münchner Kreis.

Munich, December 1980   G. Seegmüller
                                      Symposium Chairman

Inhalt/Contents

Teilnehmerverzeichnis/List of Participants

H. Afheldt, Dr.
Prognos AG, Basel (CH)

G. Arndt
Siemens AG - KZL KT 1, München

F. L. Bauer, Prof. Dr.
Institut für Informatik der TU München, München

D. Becker, Dr.
Standard Elektrik Lorenz AG - Abt. CS/FZS, Stuttgart

G. Bernau
Rank-Xerox GmbH, Düsseldorf

P. Bocker, Dr.
Siemens AG - K ZL, München

A. P. Bolle, Dr.
Dr. Neher-Laboratories, Leidschendam (NL)

G. Bolle, Dipl.-Ing.
Robert-Bosch GmbH, Stuttgart

J. L. Bordewijk, Prof. Dr.-Ing.
Technische Hogeschool Delft, Delft (NL)

A. Brandstetter, Dr.
Landeshauptstadt München, Rathaus, München

B. Czaputa,
Siemens AG, Abt. KOA 2, München

C. Detjen
Bundesverband Deutscher Zeitungsverleger, Bonn

H. Donner, Dr.
Siemens AG - DV VM PP, München

J. J. Duby, Dr.
IBM Europe, Paris (F)

K. Fischer, Dr.
Siemens AG, München

H. Forner, Dr.
IBM Deutschland GmbH - LD - SV Komm., Stuttgart

J. P. Fritz
"Die Presse", Wien (A)

H. Gabler
Fernmeldetechn. Zentralamt, Abt. T 6, Darmstadt

V. Gehrling, Dipl.-Phys.
Münchner Kreis, München

N. Georg, Dr.
Fernmeldetechn. Zentralamt, Darmstadt

F. Greiser
Süddeutscher Verlag GmbH, München

H. Grosser, Dipl.-Ing.
TeKaDe, Felten + Guilleaume Fernmeldeanlagen GmbH, Nürnberg

H. Häberle, Dr.-Ing.
DFVLR, Oberpfaffenhofen, Post Weßling

K. Haefner, Prof. Dr.
Universität Bremen, Fachbereich 2, Bremen

W. Haist, Dipl.-Ing.
Bundesministerium für das Post- und Fernmeldewesen, Bonn

H.-G. Hegering, Dr.
Leibniz-Rechenzentrum der Bayerischen Akademie der Wissenschaften, München

P.S. Heller
EDUNET, Princeton, N.J. (USA)

F. Hertweck, Prof. Dr.
Institut für Plasmaphysik der Max-Planck-Ges., Garching

H. Kabisch
Bundesministerium für das Post- und Fernmeldewesen, Bonn

W. Kaiser, Prof. Dr.-Ing.
Institut für Nachrichtenübertragung der Universität Stuttgart, Stuttgart

F. W. Kleinlein
Bundesverband der Deutschen Industrie, Köln

K. D. Kreuzer
Siemens AG, München

H. Kunze, Dipl.-Ing.
Fernmeldetechnisches Zentralamt, Darmstadt

H. Kurz, Dr.
Landeshauptstadt München, Werkreferat, München

W. Lämmle, Dipl.-Ing.
Münchner Kreis, München

H. Lenhardt
Kabelsignal, Wien (A)

H. Maier
IBM Deutschland GmbH, Abt. LD - SV,
Kommunikationssysteme, Stuttgart

H. Marko, Prof. Dr.-Ing.
Institut für Nachrichtentechnik der TU München, München

W. A. McCrum
Government of Canada, Department of Communications,Ottawa, Ontario (CND)

H. Ohnsorge, Dr.
Standard Elektrik Lorenz AG - Abt. ZT/FZ, Stuttgart

R. Oldenbourg, Dr.
R. Oldenbourg Verlag GmbH, München

H. Opderbeck, Dr.
TELENET, Vienna, Virginia (USA)

R. Pelz
Nixdorf Computer AG, Bereich CIS, München

J. Peterson, Dr.
Siemens AG - Abt. K ZL KT 1, München

K. L. Plank, Dr.-Ing.
Telefonbau und Normalzeit GmbH, Frankfurt

A. Pott
Thyssen AG, Oberhausen

E. Rahlenbeck, MA
Institut für Kommunikationswissenschaft der Universität München,
Arbeitsgemeinschaft für Kommunikationsforschung (AfK), München

E. Raubold, Dr.
Institut für Datenfernverarbeitung der GMD, Darmstadt

H. Reich, Dr.
Standard Elektrik Lorenz AG, Stuttgart

L. Rohde, Dr. Dr.-Ing.
Rohde & Schwarz, München

P. Rother
R. Hirschmann, Radiotechn. Werk, Esslingen

E. H. Schäfer
Deutsch-Atlantische Telegraphengesellschaft AG, Köln

W. Schätzler, Prälat
Zentralstelle für Medien der Deutschen Bischofskonferenz, Bonn

K.-D. Schenkel, Dr.
AEG-Telefunken - Abt. N 22 E, Backnang

W. Schleicher
Bayerisches Landesamt für Datenverarbeitung, München

E. F. Scholz
Pressevereinigung für neue Publikationsmittel e.V.,Bad Nauheim

A. O. Schorb, Prof. Dr.
Staatsinstitut für Bildungsforschung und Bildungsplanung, München

J. Schulze
Rank Xerox GmbH, Düsseldorf

G. Seegmüller, Prof. Dr.
Institut für Informatik der Universität München und Leibniz-Rechenzentrum
der Bayerischen Akademie der Wissenschaften, München

D. Solaro
Standard Elektrik Lorenz AG - Abt. Z K, Stuttgart

H. Sommer, Dr.
Deutsche Telephonwerke und Kabelindustrie AG, Berlin

K.-U. Stein, Dr.
Siemens AG - Unternehmensbereich Bauelemente, München

W. Thurl
Kathrein-Werke KG, Rosenheim

G. Tröller, Dr.
Heinrich-Herz-Institut für Nachrichtentechnik Berlin GmbH, Berlin

G. Unholzer, Dr.
Infratest Forschung GmbH, München

R. Veelken, Dr.
Siemens AG - Abtl. UB D Leitung, München

K. Vogel
Kölner Stadt-Anzeiger, Köln

H. Wahl
Siemens AG - Abt. UB Kommunikationstechnik, München

G. E. Willibald, Dr.
AEG-Telefunken, Nachrichten- und Verkehrstechnik AG,Abt. N E 4, Frankfurt

E. Witte, Prof. Dr.
Institut für Organisation der Universität München,Münchner Kreis, München

R. Zeitz
Xten Marketing, Arlington, Virginia (USA)

E. Zöpf, Dr.
Bayerisches Staatsministerium für Wirtschaft und Verkehr, München

Grußwort

A. Jaumann

München
Bayerischer Staatsminister für Wirtschaft und Verkehr

Unser privates und geschäftliches Leben wird zunehmend von Telekommunikation beeinflußt, täglich wird in Presse und Rundfunk bzw. Fernsehen
über künftige oder bereits realisierte neue Informations- und Kommunikationsdienste berichtet. Der Fortschritt auf diesem Sektor nimmt kon-
krete Formen an: Bildschirmtext wird in Düsseldorf/Neuss und in Berlin
erprobt; die Deutsche Bundespost startet einen Großversuch mit der
elektronischen Übermittlung von Briefen; die Einführung von Videotext
durch die Rundfunkanstalten ist angelaufen; der neue Fernkopier-Dienst
der Post (Telefax) bewährt sich bereits und für die Mitte der 80er
Jahre ist die Ausstrahlung regelmäßiger Rundfunk- und Fernsehprogramme
mit Hilfe geostationärer Satelliten zu erwarten. Nicht zu vergessen
auch die vielfältigen Möglichkeiten der dezentralen Textbe- und verarbeitung mit neuen Büro- und Verwaltungskommunikationssystemen. Die
schon beinahe lawinenartige Veränderung unserer Arbeitswelt und auch
unserer privaten Lebensgewohnheiten wurde durch den rapiden Fortschritt
auf dem Gebiet der Mikroelektronik ausgelöst. Ein Ende dieser Entwicklung ist nicht erkennbar. Auch die Industrie selbst ist überrascht, wie
schnell und qualitativ hochwertig sich die Glasfasertechnik entwickelt.
Und aus Japan ist kürzlich bekannt geworden, daß eine 50-fache Erhöhung
der Leistungsfähigkeit eines Halbleiterchips, jenes elementaren Bau-
steins der Computertechnik, gelungen scheint.

Die wirtschaftliche Bedeutung dieser Vorgänge und damit auch das
Interesse des Wirtschaftspolitikers an den neuen Informationstechniken
liegt auf der Hand. Es scheint, daß wir es hier mit der Basisinnovation
für die 80er Jahre zu tun haben, glücklicherweise mit einer Technologie,
die einen relativ geringen Rohstoff- und Energiebedarf hat. Ich bin
überzeugt, daß die zukünftige Informations-Infrastruktur eine enorme
Triebkraft für einen besseren Austausch von Wissen und Meinungen zwischen den verschiedensten Wirtschafts- und Verwaltungsorganisationen
national und international sein wird. Es ist zu hoffen und es muß auch
wirtschaftspolitisch alles getan werden, daß insbesondere auch die mit-
telständische Wirtschaft an den Vorteilen eines effektiveren Technolo-
gietransfers teilnehmen kann.

Nicht minder wichtig, wenn nicht sogar von noch größerer Tragweite
sind die aus der Einführung neuer Informationssysteme erwachsenden ge-
sellschaftspolitischen Aspekte. Die Frage ist gewiß nicht leicht zu be-
antworten, welchen Nutzen der Mensch insgesamt aus der Einführung solcher Neuerungen ziehen kann. Es ist wohl schwer vorherzusagen, in wel-
cher Weise sich dadurch sein Wohlbefinden, etwa in Form größerer Frei-
zügigkeit oder verbesserter Möglichkeiten der Selbstverwirklichung in
der eigenen Arbeits- und Privatsphäre, erhöhen läßt. Die Untersuchung
einer so komplexen und schwierigen Thematik von Wissenschaftlern einer
Fachrichtung allein und ohne entsprechende Voraussetzungen für prakti-
sche Versuche (Pilotprojekte) dürfte nicht möglich sein. Hierin sind
vielmehr Experten der verschiedensten Disziplinen, z.B. aus dem Kreise
von Naturwissenschaft und Technik, der Wirtschafts- und Sozialwissen-
schaften, dem Medienbereich, aber auch aus dem Kreis der Wirkungsfor-
schung aufgerufen, interdisziplinär und vorurteilslos zusammenzuarbei-
ten. Angesichts der heute oft mit viel Emotion vorgebrachten Argumente
gegen die neuen Medientechnologien sind wissenschaftlich fundierte und
gesicherte Grundlagen und Erkenntnisse zu erarbeiten und - was mir besonders wichtig erscheint - auch allgemein verständlich darzustellen,
um somit auch in der breiten Bevölkerung ein Gegengewicht gegen den
heute so modern empfundenen Zeitgeist "wider die Technik" zu schaffen.

Besonders der Politiker braucht die sachlich fundierte Erkenntnis,
um bei den anstehenden Entscheidungen die letztlich immer verblei-
benden Unwägbarkeiten so gering wie möglich halten zu können.
Die Problemstellung ist international; wir brauchen auch die Erfahrungen und Erkenntnisse des Auslands und müssen umgekehrt zum Er-
fahrungsaustausch bereit sein. Ein Blick auf das Programm dieser
Fachkonferenz zeigt, daß es dem Münchner Kreis auch diesmal wieder
gelungen ist, entsprechend seiner interdisziplinären und übernatio-
nalen Zielsetzung hervorragende Fachleute für diese Fachkonferenz
als Referenten zu gewinnen.

Ich darf Sie alle, besonders die Gäste aus Übersee, die einige
Strapazen auf sich genommen haben, um zu uns zu kommen und hier
über ihre Erfahrungen zu berichten, recht herzlich willkommen
heißen und für Sie hoffen, daß Sie vielleicht auch etwas Zeit finden, um Land und Leute bei uns kennenzulernen. Ich wünsche der
Konferenz einen erfolgreichen Verlauf.

Functions and Realizations of Value-Added Networks

S. Heller
Princeton, N.J./USA

Summary

EDUNET is an international facilitating network for computing in
higher education and research. The communications technology underlying Edunet is TELENET and TYMNET. The different ways in which a
customer host may be interconnected with the networks are explained.
The costs involved in using telecommunications facilities in the
United States are discussed. Finally the user support of EDUNET is
described.

Zusammenfassung

EDUNET ist ein internationales anwenderorientiertes Netz für Lehre
und Forschung. Es setzt auf den Transportnetzen TELENET und TYMNET
auf. Zunächst werden die unterschiedlichen Netzzugangsmöglichkeiten
erörtert. Dann werden die Gebührenstrukturen für die Inanspruchnahme
von Telekommunikationsdiensten in den Vereinigten Staaten diskutiert.
Schließlich werden die Leistungen von EDUNET beschrieben.

In my remarks today I would like to present a brief overview
of the functions and realizations of value-added networks as they
exist in the United States today. My observations are based
primarily on my experiences with Telenet and TYMNET, the two
dominant public packet-switched data communications carriers now
operating in the U.S. My organization, EDUNET, is a large
customer of Telenet and several of our member universities are
customers of TYMNET. Our combined usage of these carriers exceeds
$600,000 per year. I am therefore directly concerned with the
quality and cost of services provided by these carriers.

Let me begin by giving an example of one of the prevalent
methods of using Telenet or TYMNET. In the diagram (see Figure 1)
it is convenient to think of the public networks as very large
black boxes with many nodes on the edges. These nodes are the
points at which host computers and terminal users connect to the
network. Telenet and TYMNET each have between 100 and 200 such
nodes. The communication link between the network and the
customer site where a host computer is located is a line leased
from the telephone company typically operating at 2400 – 9600 bits
per second. This line is connected to a multiplexer/controller

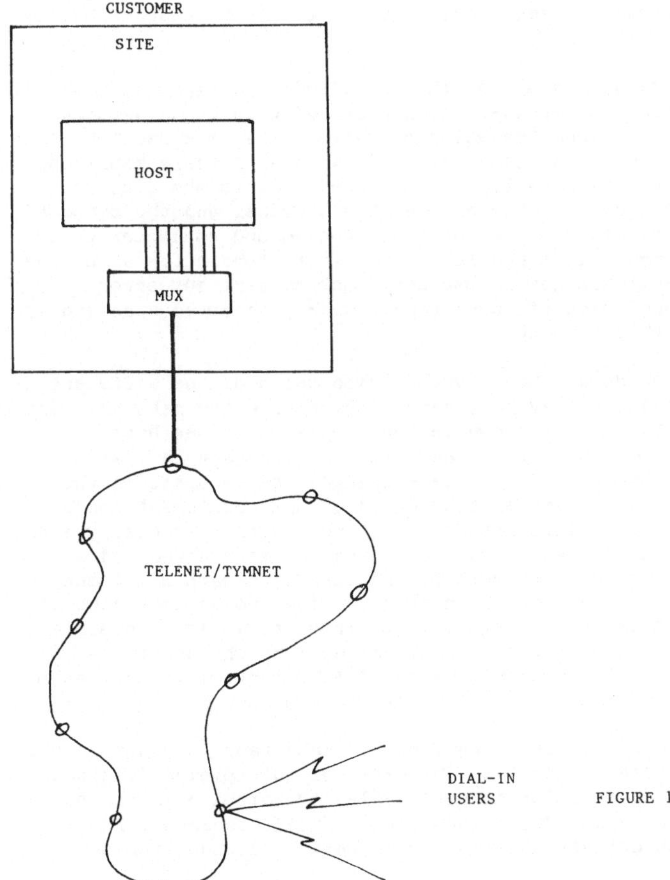

FIGURE 1

rented from the network but physically located near the customer's
host computer. The connections between the multiplexer and host
are usually a number of asychronous lines operating at 110 - 1200
bits per second. From the point of view of the host, these lines
appear to be local hardwired terminals.

At the user end, most traffic is handled using the ordinary
dial-up voice telephone network. Virtually every low speed
asychronous terminal that operates at speeds between 110 - 1200
bits per second is compatible with Telenet and TYMNET. This is a
very important benefit.

The networks provide speed, code, and limited protocol
conversion. Speed conversion means that the instantaneous
transmission rate into the network need not match the
instantaneous rate out of the network. Of course, since the
networks buffer only a small amount of data, differences in
instantaneous rates can be tolerated only for short periods of
time unless some form of flow control is available at the host
and/or terminal. Code conversion in the network allows, for
example, an EBCDIC terminal to interact with a host that supports
only ASCII devices. Very limited forms of protocol conversion
handle such functions as automatic insertion of line feeds and
null characters to prevent data loss during a carriage return on a
printing terminal. The networks do not support conversion for
other functions such as cursor positioning or forms handling.
These conversions, if required, continue to be the responsibility
of the host.

Another major feature of the value-added carriers is the full
error detection and retransmission provided within the network.
The goal of error-free transmission between the host and the
network and within the network itself has been largely achieved.
However, one error-prone link remains and that is the dial-up
asychronous connection between the user terminal and the network.
Noisy telephone lines are a constant problem and character errors
are not uncommon. This problem is likely to remain for some time
since most terminals now in use could not support any error
control protocol even if there were a widely accepted standard
protocol for this purpose.

Let me now return to a further discussion of the different
ways in which a host may be interconnected with the network. The
use of a multiplexer as shown in the diagram provides both
advantages and constraints. The dominant advantage is that
absolutely no hardware or software changes are required in the
host. Consequently, any size or brand of host equipment can be
readily connected. This method of interconnection is costly since
the multiplexing hardware must be rented and maintained and a
hardware port on the host must be dedicated for each simultaneous
user to be served through the network. Since no network accouting
information is passed through the multiplexer, it is impossible
for the host to know the origin of the call or whether it is
prepaid or collect. The only means of flow control is the rather
primitive use of the X-ON/X-OFF paper tape protocol.

A much more efficient and flexible interface is shown in the
next diagram (see Figure 2). The external multiplexer disappears.
The multiplexing function is now handled by software in the host
or a host front end. The international CCITT standard X.25
protocol which defines the connection between a computing device

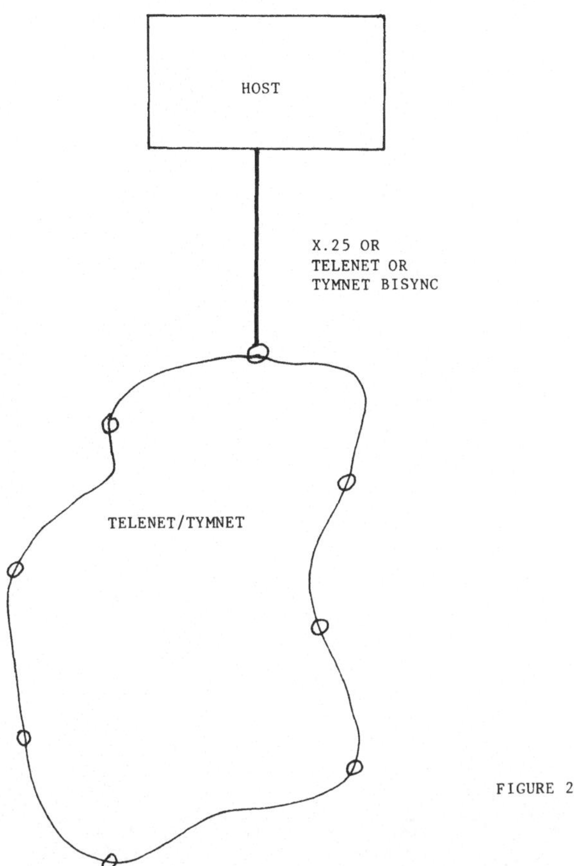

FIGURE 2

and a packet-switched network is now supported by both Telenet and
TYMNET for this type of connection. Full flow control as well as
error detection and retransmission are integral to X.25.
Information about the origin of the call is available. Since the
host software to support X.25 is available only for a few types of
hosts, customers may still find it necessary to develop and
support their own software for an X.25 interface. Since many
customers are reluctant to modify operating system software, only
a small fraction of the host connections are based on X.25. The
only major computer system vendors in the United States that
support X.25 are Prime and Data General. Fortunately, other major
vendors have given clear indications that they will eventually
support X.25. Digital Equipment Corporation recently announced
that X.25 would be supported in Phase III of Decnet. However, it
may be twelve months before the software is available to customers
in the U.S. In the case of IBM, Lewis Branscomb, Chief Scientist
recently announced that X.25 support should be made available for
selected IBM products to supplement the functions provided by
IBM's Systems Network Architecture. It's ironic to note that X.25
support from IBM for certain products is already available in
Canada, France, the Netherlands, and here in Germany. I and many
others eagerly await the announcement of its availability in the
United States.

I should note that there is another set of protocols that
were developed by Telenet and TYMNET independently before X.25 was
defined. The early Telenet protocol is structurally similar to
X.25 and is in use by several customers. Telenet customers that

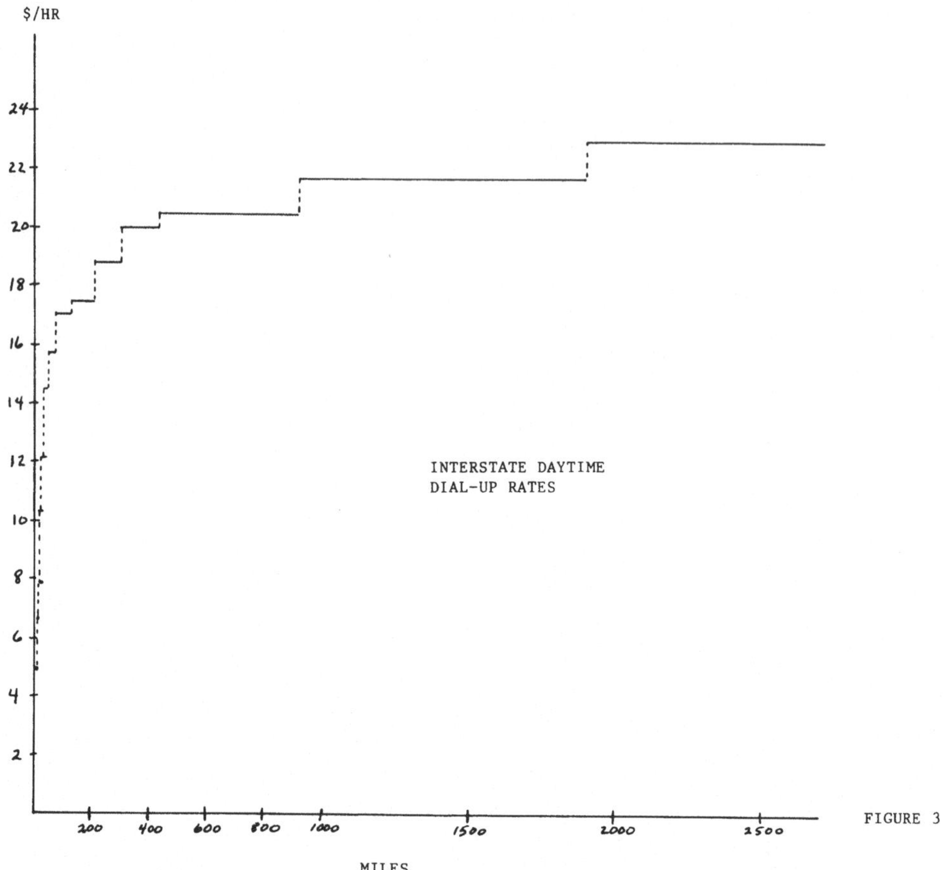

FIGURE 3

have IBM equipment with a 3705 front end may lease the Telenet
protocol software from Telenet. Within a year, this early
protocol will be replaced by standard X.25.

TYMNET supports a binary synchronous protocol which operates
on IBM 3705 front ends using TYMNET supplied software. TYMNET
also supplies software for a few other hosts. These non-standard
protocols are likely to disappear over time to be replaced by
standard X.25.

It is interesting to note that most software interfaces now
in use support only incoming traffic. The reason for this is that
most current network customers are time-sharing service bureaus
that wish to provide inexpensive terminal access to their
employees or customers. Direct host-to-host traffic is not very
prevalent on public networks. This type of traffic requires
higher level protocols to provide, for example, file transfer
between hosts. Currently, there are no national or international
standards for file transfer over a public network.

I would now like to look at some of the costs involved in
using telecommunications facilities in the United States. As a
point of reference, it is interesting to look at the cost of using
the public dial telephone network. The chart (see Figure 3) shows
the hourly cost of a daytime call as a function of distance.
These costs are regulated by the Federal Communications Commission
and apply to all calls that cross a state boundary. The rates
have not changed during the past five years. Between 1,000 and

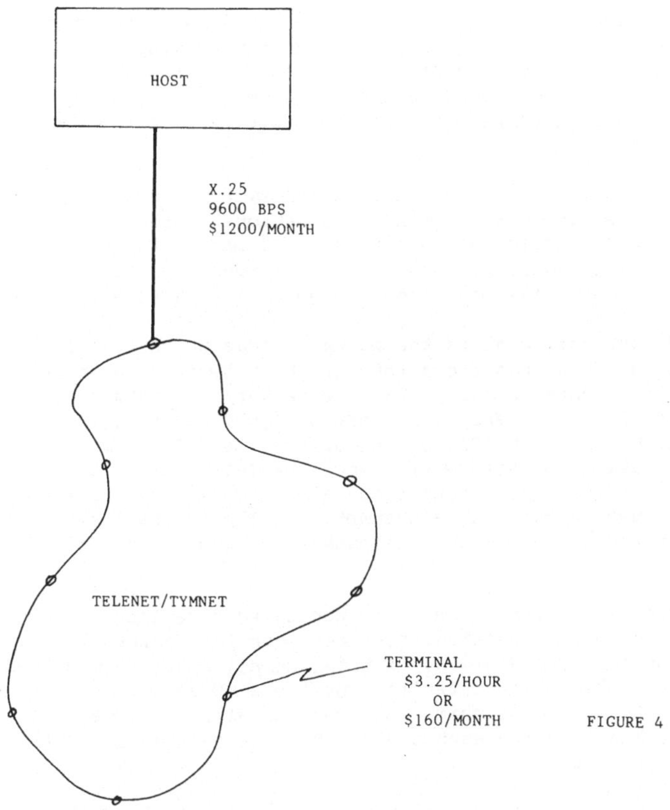

3,000 miles, the cost is almost independent of distance. A
characteristic of both Telenet and TYMNET is that the cost is
totally insensitive to distance. In Canada, the cost of Datapac
is dependent on distance. For dialed calls that do not cross a
state boundary, the rates are regulated by the state.
Consequently there are fifty such charts. In my home state, New
Jersey, there is no such thing as a free call. The cheapest
possible call is $.84 per hour. A call of 30 miles costs $6 per
hour. For a dial-up terminal user, these costs are very
significant. They must be added to the costs associated with the
network and may, in fact, dominate the network charges.

Now let's look at some of the costs of using Telenet in a
service bureau application. This would be an application in which
a host provides low-speed terminal service to a large number of
customers widely scattered throughout the country. The diagram
(see Figure 4) shows the configuration. The host uses the X.25
protocol and is linked to the network through a 9600 bit per
second line leased from the telephone company by the network. The
typical cost for the line, modems, and the X.25 network port is
$1,200 per month. The host-resident X.25 software, if available,
would cost $500-1,000 per month.

The number of simultaneous terminals this link could support
depends, of course, on the speed of the terminals and the nature

of the application. In typical interactive applications where the
operator spends time thinking, keyboarding, and reading, the link
could support 50-75 terminals at 30 characters per second or 20-40
terminals at 120 characters per second. Based on reasonable
assumptions regarding usage patterns, the cost per terminal hour
would be well under $1.

Now let's look at the costs associated with the dial-up
connection between the user terminal and network. The network
charges $3.25 per hour for dial-up usage. If usage at a given
location exceeds 50 hours per month, a less expensive alternative
is leasing a private dial port from the network at $160 per month.

A third cost component is the packet charge which is related
to the amount of data that flows through the network. In Telenet
the basic unit of data is the packet. A packet can contain up to
1,024 bits or 128 characters. The cost of 1,000 packets, which
could conceivably contain 128,000 characters, is $.50. However,
in this application, packets would rarely be full. Experience
suggests that, on average, packets contain only 15-20 characters
each. Again, making reasonable assumptions, the interactive
terminal user would consume one kilopacket per hour at a cost of
$.50.

To summarize the three cost components, we have less than $1
per hour at the host connection, $.50 per hour for packets, and
$3.25 per hour for user dial-in. A total hourly cost of less than
$4.75 per hour. The Telenet tariff contains a number of
provisions that can reduce this cost substantially. Volume
discounts, bulk port arrangements and other options can cut this
cost by half or more.

TYMNET has a cost structure that is very similar to that of
Telenet. One significant difference is that TYMNET charges by the
character rather than by the packet. In the terminal to host
time-sharing application, the Telenet and TYMNET traffic charges
are about the same. If you were able to fill the Telenet packets,
Telenet would be much cheaper.

The weakest and most expensive link between the user and the
host is the dial-up connection. One way to improve this link and
reduce this cost is shown in the next diagram (see Figure 5). If
there is a sufficiently large user community at one site, it will
be cost-effective to install a multiplexing device which can
interface with the network using X.25. If the multiplexer and
terminals are in the same building, they can be inexpensively
wired together.

A next step in this same direction is to replace the
multiplexer with a host computer (see Figure 6). The host can
provide some processing power useful to the local users and an
efficient connection to the distant host for processing functions
or access to databases not available on the local host. The most
serious barrier to widespread adoption of this approach is the
nearly complete lack of software to support this form of
distributed processing across a packet-switched network.

Taking one further step in this direction (see Figure 7), we
come to a new technology for local networking based on Ethernet,
developed by Xerox. Ethernet is based on a passive medium
consisting of coaxial cable. It is cheap, and so simple and

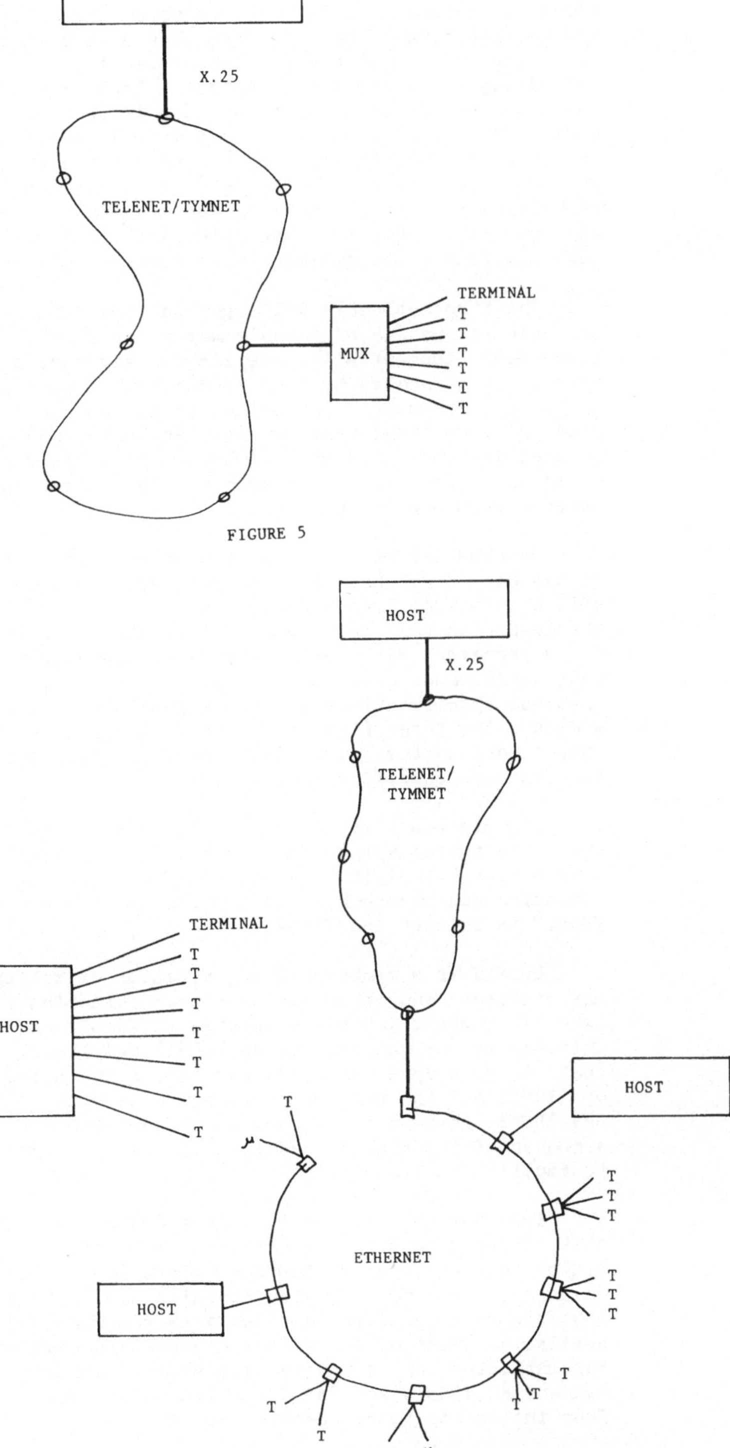

FIGURE 5

FIGURE 6

FIGURE 7

reliable, that it should never break down. Each terminal, microprocessor, large system, or other device is connected to the cable through a transceiver. Each transceiver has a unique 48 bit address. Information is transferred on the cable in packets which include the data to be sent, the addresses of both the sending and receiving unit, a 16 bit protocol type field, and a 32 bit cyclic redundancy check for error control. Each transceiver monitors the cable before transmitting to be sure it is clear and it monitors during transmission to detect any interference during transmission. If there is interference, the packet is sent again when the cable is clear. To receive data, each transceiver recognizes its own address. It accepts packets with that address and ignores all others. The receiving station sends the source an acknowledgement indicating it has received the packet.

The Ethernet cable will operate at a data rate of ten million bits per second and each cable may be as long as two kilometers. Since each Ethernet will have its own address, a transceiver may be used as a gateway to link several Ethernets together. This technology looks very promising for interconnecting different kinds of computers, peripherals, terminals, and office equipment located in a building or a closely spaced cluster of several buildings. Several of our member universities have experimental Ethernets in operation.

Perhaps the most unusual characteristic of Ethernet is that it has been developed over the past year through a cooperative effort involving Digital Equipment Corporation, Xerox, and Intel, the leader in very large scale integrated circuits. These three firms represent different but very complementary segments of the computer/communications industry. Electrical, logical, and protocol specifications are to be published in the next three months. The three firms have encouraged others in the industry to participate in the development and to develop products based on the Ethernet specifications.

I would now like to spend a few minutes describing EDUNET, the activity for which I am responsible. EDUNET is an international facilitating network for computing in higher education and research. The communications technology underlying EDUNET is Telenet and TYMNET.

EDUNET is a membership organization of colleges, universities, and other non-profit organizations. We currently have 110 members including several in Canada and one in Europe. Thirteen of the members act as EDUNET suppliers. One or more of their large campus computing facilities are connected to Telenet or TYMNET and the many programs on these systems are available to the EDUNET membership by remote terminal access. EDUNET provides a number of information, administrative, and user support services to facilitate this access.

Information services include a comprehensive member's guide that contains detailed information about the supplier's hardware, system software, rates, hours of operation, logon procedures, documentation, etc. It also contains abstracts and sample sessions of many of the applications programs and data bases available. Most of our effort has been focused on the unique, sophisticated, or otherwise interesting resources rather the mundane FORTRAN's and BASIC's which would usually be available from the user's local computer center.

Another set of information services is designed to locate a program that will meet a specific user need. An online keyword searchable catalog of programs is maintained under the SPIRES system at Stanford University. More than 600 programs are catalogued with an abstract, documentation list, level of support, access instructions, contact person, and other descriptors that enable the user to determine whether the program will meet the need. Other means for locating desired applications programs include broadcast electronic mail messages and periodic publication of user needs.

A primary administrative service is the complete account setup and billing service maintained by the central network staff. There is very little delay or red tape in opening one or many accounts at one or more suppliers. Full details of network usage are available to the member in the form of a consolidated monthly bill with supporting detail for each account.

User support includes access by telephone or electronic mail (using the EDUMAIL terminal-based message system) to network consultants who can handle most problems that users encounter. Queries not answered by network consultants are forwarded to the supplier or developer of the program in question. Special documentation has been developed to answer the most frequently asked questions such as how to delete a character or suspend the printing of output.

The thirteen EDUNET suppliers include:

> Carnegie-Mellon University
> Cornell University
> University of Delaware
> Massachusetts Institute of Technology
> University of Minnesota
> New Jersey Institute of Technology
> North Carolina Educational Computing Service/
>   Triangle Universities Computation Center
> University of Notre Dame
> Rice University
> University of Southern California
> Stanford University
> University of Wisconsin-Madison
> Yale University

All of the suppliers are connected to Telenet and three are also connected to TYMNET.

A product recently introduced by EDUNET is a microcomputer based software system called EASy (see Figure 8). This system is written in Pascal and operates on an Apple microcomputer. This system provides two very important benefits for EDUNET users: simplified access and substantially reduced cost.

First, it greatly simplifies the process of accessing a remote host by automatically generating the sequence of commands and parameters required by the network and host. A typical sequence might include a speed recognition character, a terminal identifier, a login command, an account number, a password, a personal identifier, and a run time limit. This sequence may be fifty keystrokes and needs to be entered perfectly. Users find this tedious and frustrating. The EASy system allows the user to

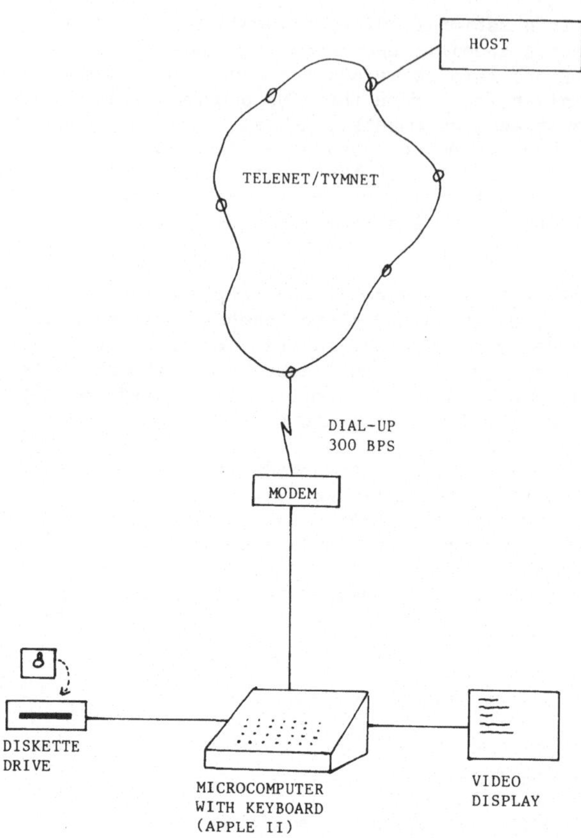

FIGURE 8

store this entire sequence on a "floppy" diskette. It may then be
played back with one keystroke. This automatic login feature is
especially attractive for the user who interacts with several
different hosts.

A second major benefit of the system is its ability to
automatically transfer files in both directions between the
microcomputer diskette and the remote host. We wanted to
implement a file transfer protocol that would require no changes
to the host front end or the operating system. The host
transmitter-receiver programs would have to operate as
applications-level programs. The communications channel would
ordinarily be a dial-up call at 300 bits per second to the packet-
switched network to which the host was connected. Some form of
error detection was desired. Flow control was also useful since
congestion on the host results in widely varying response times.
Transmission would either have to be prompted or long pauses would
need to be present to insure that the host was ready for the next
block of data.

Two protocols have been implemented. The simplest relies on
a paper tape protocol using X-ON/X-OFF for flow control. The
simplest version has no error detection. The second protocol uses
the same flow control scheme but has a simple block checksum for
error detection. The block is retransmitted if an error is
detected. This second scheme has worked quite well for
transferring text, numeric data, and program source code files.
Files containing certain control characters cannot be sent since
the host would interpret them and perform unintended functions
such as line delete. This would destroy the integrity of the
file.

Electronic mail is an application which can take good
advantage of the file transfer capability of EASy. The Apple
microcomputer has an excellent screen-oriented editor which is
very convenient for preparing text messages. Messages are
prepared and edited off-line and stored on the diskette so there
are no charges for use of the network or host. The user then
invokes the automatic login facility to establish the connection
with the central mail site. The user enters the name of the
addressee and then initiates the file transfer facility to
transfer the message on the diskette to the host.

It is interesting to compare the economics of electronic mail
using a conventional terminal and using the EASy system. Consider
a message that is one full typed page, about 3000 characters. At
normal typing speed and allowing some time for correction of
errors, it would take about 20 minutes to enter. If this were
done on a conventional terminal connected to a remote host, this
would result in a connect time of 20 minutes. Based on the prices
of the host we use for mail, this would cost about $5. Using the
file transfer capability at 30 characters per second, which is
about ten times faster than typing speed, the connect time would
be about 2 minutes at a cost of $.50. This ten to one savings is
rather dramatic.

This concludes my remarks on the functions and realizations
of value-added networks in the United States. I would be happy to
answer any questions.

DISCUSSION after the lecture of Paul S. Heller

Question: The expression "Value Added Network" rises some
interest in Europe because we asked ourselves whether our
networks in Europe fulfil the requirements of the users
better or worse than these value added networks. Now having
listened to your contribution I learnt that the expression
VAN in your opinion only denotes a simple packet switched
network where the added value is more or less related to
a more reasonable tariff structure, which cannot be
provided by a circuit switched network. Is this understanding correct?
Does that sort of VAN really match your wishes concerning
a multipurpose data network? Is that an aid to come to an
open system, a question which is of great interest
especially for universities?

Answer: Now, first to the question what a VAN, a Value Added
Network, is in my way of thinking.
I don't think of a packet switched network as being simple.
Compared with a circuit switched one there is a lot of
added value, e.g., error control, flow control, conversions
of code and formats, speed adaption etc. The clear benefit
to us operating primarily in a terminal-to-host environment
is reduction of cost. From a technical point of view we can
do what we are doing now without a VAN except we would pay
about five times as much. The result is that we wouldn't do
what we are doing now. Looking towards the future, towards
some non-economic characteristics we would like to go into
the direction of some Ethernet-like sort of interconnections
in the university areas within USA and hopefully on a world-
wide-basis. From a hierarchy view of networks a local network based on Ethernet-, circuit switched-, time division
multiplexed, or whatever technology, should be interconnected
to a regional network, a national network, and an international network. To my extent as a user I don't particularly

care whether such interconnections are satellite based or
what the underlying backbone structure is. I would like to
be able to have protocols that are reasonably standard
which we can encourage our universities to adopt for the
full range of applications including file transfer, inter-
process communication, standards for message exchange and
formats etc.

Question: The remaining question is what are your requirements and
your intent to obtain networks which make open systems
easier?

Answer: I think the reason I mentioned the Ethernet architecture
is the design goal - I think along with most of the packet
networks - to provide open system capability. In order to
realistically achieve an open system you need to have
standards and I think what is really the single most
important barrier to have an easier interconnection of
different types of equipment is standards. X.25 provides
only a standard for one level presuming not the level of
an office manager wanting to connect a real processing
device which may involve file transfer. I am not sure
whether standards can really come before there is some
great experience with these applications. Because standards
are often developed too quickly and ignore several
suggestions you often find you can't do things within the
constraints of a standard. Just this is the criticism of
X.25 by some of the very leading managers of networking
technology who talk about packetizing voice, real time
applications or process-to-process applications. I think
this criticism concerning X.25 is probably right but it
was unknown by the time X.25 was defined.

Question: What role does ARPA play now in this picture you designed?

Answer: That turns out to be a great current question. Much of the
           development of the ARPA-Network was promoted and managed
           by Mike Roberts who was the founder and continuous president
           of TELENET. So the experiences gathered with ARPANET
           directly contributed to the formation and current operation
           of TELENET. The position it plays currently in the research
           community is that it is heavily used by a fairly small
           segment of the computer science and artificial intelligence
           and communications research communities, because policies
           of the U.S. Department of Defense is not generally available
           for use by wider groups of researchers. For that reason there
           have been some recent discussions involving ARPA and the
           National Science Foundation in the U.S. and several computer
           science research groups to try to identify a way which could
           be more generally available than the ARPANET and would prob-
           ably not be under the constraint of being available only
           under a defense related contract. It is likely that the
           facilities of some networks continue to be needed so that
           some of the non-standard networks will survive but not
           jeopardize a production facility on which you wish to rely.

Question: What kind of protocols do you use above X.25? Are they
           the same for different hosts or not? Are they provided by
           your company if the customer has selected a certain kind
           of host?

Answer: Most sites use micro-coded interfaces based on the
           characteristics of the standard X.28, defining asynchronous
           terminal characteristics. We currently do not rely upon
           particular standards for file transfer. The micro-computer-
           based system uses some very ad-hoc standards because we are
           operating in an environment that we can't pick up existing
           standards such as the ARPANET FTP. There are several carriers
           working on file transfer standards, but I am not sure which
           of them will get in favour by the Standards Organization.

Question: To what extent is ETHERNET an Open System? I have learnt
that Ethernet is announced now merely as a XEROX product.

Answer: Since the specifications of ETHERNET haven't been at
least publicly announced Stanford University which
is working with it, the XEROX people which are working
independently on the Ethernet concept, the outcome will
be structurally compatible with most of the equipment
they would like to attach to it e.g. text processing
equipment, realtime data collection equipment. So, in a
sense, if you worry about an interface standard it is an
open system. But if you want to buy that interface there
is the task to adapt the particular protocol of the net.

Question: Are your customers happy with your very slow and not
very comfortable terminals, the asynchronous 300 baud
teletypes? In Germany the usage of teletype like terminals
is not so wide spread. Most people are connected directly
to their local computer and they are accustomed to the
manufacturer provided terminals which allow for stream
oriented, context oriented work, the terminals having
high speed and low error rate. Now that packet switching
is introduced in Germany, X.29/X.28 and pad oriented
services are available. This seems to be a step back
for our pampered users. How do your customers react?

Answer: Well, our customers are not yet accustomed to full screen
high bandwidth and find it fine. That is a service which
ist not present locally. Those who are familiar with full
screen devices like people in the ARPANET for example do
not tolerate a line-by-line 30 character per second device.
But this doesn't play that role in a university's
environment because the costs for the human operator, a
student, are nearly up to zero. I believe that the standards
X.25/X.28/X.29 do not impose any constraints regarding the
bandwidth, at least not in the range up to 19.2 kbit.

Question: We have a worldwide telephone network and a worldwide
          telex network. These networks may be a little bit
          oldfashioned, but they are beautiful because everybody
          can call everybody and communicate to everybody without
          any difficulties. So my question is: Do there exist
          interfaces and crossings between the different
          communication networks like for instance EDUNET,
          TELENET and the others, since I don't see what the
          use of a network is if you can't reach another one and
          have an actual link to it. So if these networks are
          good only for universitiees this would be of rather
          little interest and limited practical application for
          the public.

Answer: I agree. I would very much like all networks linked
          together and access points in every home, perhaps cable
          television circuitry via satellite with access in every
          home so that everybody is connected with everybody.
          But that's too early now, that's for the future.

New Developments in Packet Communication

Opderbeck
Vienna, Va/USA

Summary

The TELENET network is presented. The expanded terminal support
is based on the CCITT approved X25 interface. All necessary
protocol conversion is performed by a dedicated multi-microprocessor. The long distance communication takes place via
satellites. The impact of higher transmission rate and larger
delay is discussed. The packet radio facilities and local area
bus systems for the local distribution of traffic are introduced.
The electronic mail service TELEMAIL is described as an
application of TELENET.

Zusammenfassung

Das TELENET-Netz wird vorgestellt. Die Unterstützung verschiedener
Terminaltypen setzt auf der einheitlichen X.25-Schnittstelle auf,
wobei alle erforderlichen Protokoll-Umwandlungen von einem speziel-
lem Multi-Mikroprozessor vorgenommen werden. Die Übertragung über
weite Strecken bedient sich der Satellitentechnik. Auf die Auswirkung von höherer Übertragungsrate und größerer Verzögerung wird
eingegangen. Geschildert werden auch die für die lokale Datenver-
teilung eingesetzten Techniken wie Funk und Bussysteme. Als ein
auf TELENET aufbauender Dienst wird TELEMAIL, eine elektronische
Briefvermittlung, beschrieben.

1. THE TELENET NETWORK

The Telenet packet-switched public network consists of eight major
switching centers (Class I offices) and a large number of smaller
offices (Class II offices) which provide dial-in access to the network
(Figure 1). For the rest of 1980 and 1981 a very aggressive expansion
program is planned. The total number of Class II offices will be
increased to about 250. At least two more Class I offices will be
added. Also, beginning in 1981 Class II offices will be directly
cross-connected.

Figure 2 shows the Telenet local access facilities. The most common
usage is the Telco local dial-in facilities. The network also allows
users to dial-out over Telco facilities as well as access terminals
over leased lines. Multiple connections are supported via data PBXs
over leased lines. The plans for packet radio access facilities shown
in Figure 2 will be discussed in a later section.

The only mode in which the packet-switched communication subnet
interacts with the outside is via the CCITT approved X25 interface.
All non-X25 terminals require a protocol conversion process. The
protocol conversion modules form a ring around the basic X25 subnet
(See Figure 3). Some of the protocol conversions currently supported
by Telenet are used with asynchronous ASCII or EBCD terminals,
2780/3780 batch terminals, HASP workstations, and 3270 display
terminals.

The term value-added network is derived from the value these protocol
conversion modules add to the basic transport mechanism. Packet-
switching in and by itself for the purpose of replacing leased lines
does not create a value-added network. The ability to easily convert
protocols, in particular the ability to do speed and code conversion,
is, however, heavily dependent on an efficient interface into the
communications subnet.

The hardware base that supports the packet-switching subnet as well as
the protocol conversion processes consists of a multi-microprocessor.
This system, designated the Telenet Processor (TP) series, is based on
an expanded set of microprocessors. It has been designed from the
beginning for high efficiency in communications applications and no
concessions have been made to other applications. Processing power and
memory requirements can be matched to the number of lines to be
supported up to maximum of more than 250 lines. Line interfaces are
available for asynchronous, bisynchronous, and HDLC-type communications
protocols.

PACKET COMMUNICATIONS OVER SATELLITES

So far commercial packet-switched networks have been limited to
communication over mostly terrestrial lines. An interesting new mode
of packet communication is available by virtue of the broadcast
capability of satellites. Figure 4 shows the conceptual organization
of a packet satellite facility serving four earth stations.

THE TELENET NETWORK

FIGURE I

TELENET LOCAL ACCESS FACILITIES

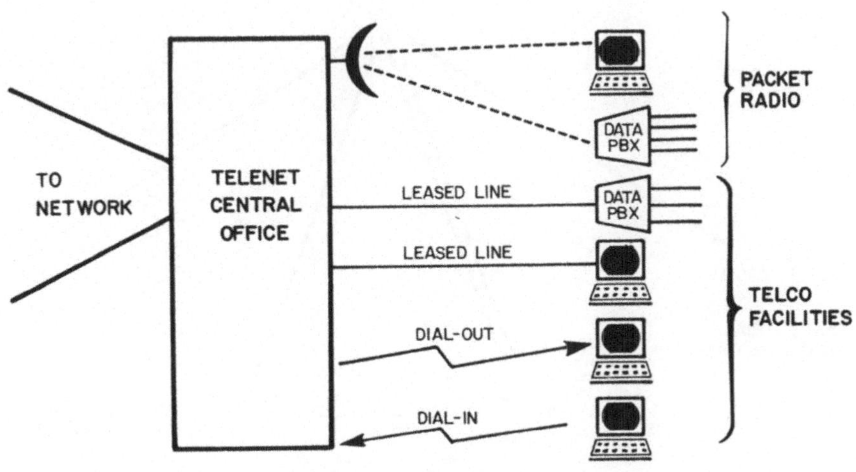

FIGURE 2

EXPANDED TERMINAL SUPPORT

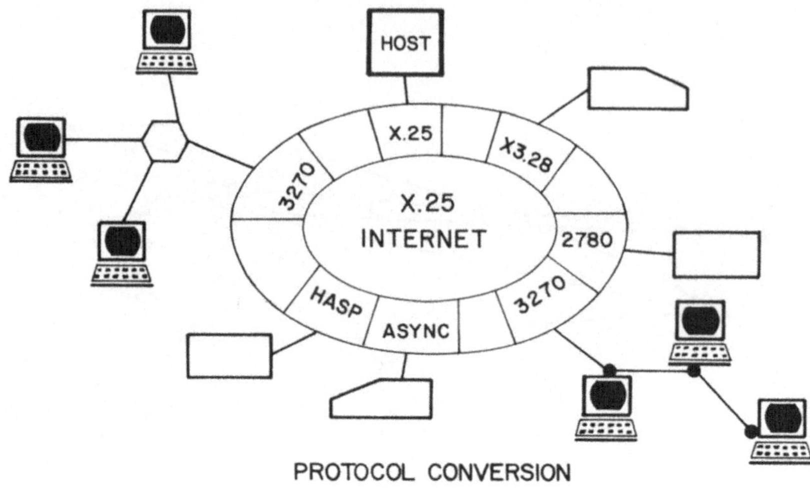

PROTOCOL CONVERSION

FIGURE 3

BROADCAST TRANSMISSION MODE

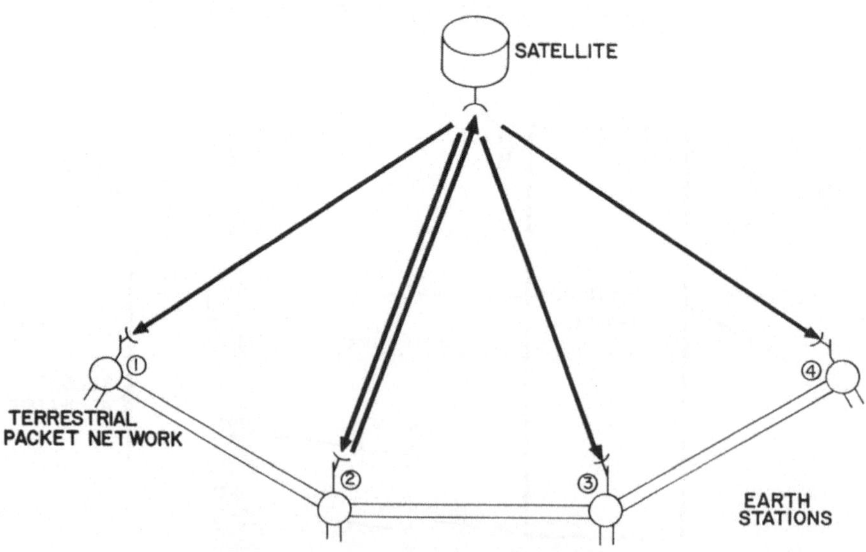

SATELLITE PACKET SYSTEM

FIGURE 4

There are two major considerations for satellite systems which distinguish them from pure terrestrial transmission systems: 1. The higher speed (megabits instead of kilobits) and 2. the larger delay (275 milliseconds one-way delay). The higher speed of the satellite circuit requires new hardware interfaces which provide the software with more capabilities. The longer delay can cause considerable problems with interactive-type connections where response times of less than one second are often required. For batch-type transmissions, on the other hand, the high throughput that satellites provide can relieve the terrestrial network of a lot of traffic.

Traditionally, satellite circuits have been used by dedicating transmission bandwidth to any two earth stations that needed to directly communicate with each other. For a large number of earth stations this allocation scheme can lead to a serious fragmentation of transmission bandwidth and therefore to inefficient usage. In the case of packet communication over satellite, transmission bandwidth is allocated on a per node basis. This is possible because the packet approach makes use of the broadcast nature of satellite transmissions. Every packet a station sends is not just received by the station that it is intended to go to but by all stations. Each station examines all the packets received on the downlink from the satellite and discards those that are not addressed to itself. The problem with this approach is that the scanning of packets needs to be done at the high speed of the satellite link and therefore requires specialized hardware.

The assignment of transmission bandwidth to the stations is done by a master station according to the current traffic requirements. It is possible to dynamically change this assignment. The stations need to be synchronized properly so no transmission from one station will overlay the transmissions from other stations.

PACKET RADIO

Packet radio facilities represent an alternative to leased lines for the local distribution of traffic. Figure 5 shows a packet radio central node connected to the terrestrial packet network. The central node communicates with the packet radio station via microwave transmission facilities. Each packet radio station can be viewed as a device that allows terminals or host computers access to the packet network.

The central station is equipped with an omnidirectional antenna, i.e. a packet transmitted by the central station is normally received by all packet radio stations. Similar to the use of satellite facilities discussed before, the packet radio station scans all incoming packets and selects those for further processing that are addressed to itself.

There is no contention for transmission bandwidth for the outgoing link from the central station to the packet radio stations because there is only one transmitter involved. This is not true for the traffic generated by the terminals and host computers connected to the packet radio stations. The traffic flowing into the central station needs to be controlled carefully to avoid excessive overlap of transmissions. A

PACKET RADIO SYSTEM

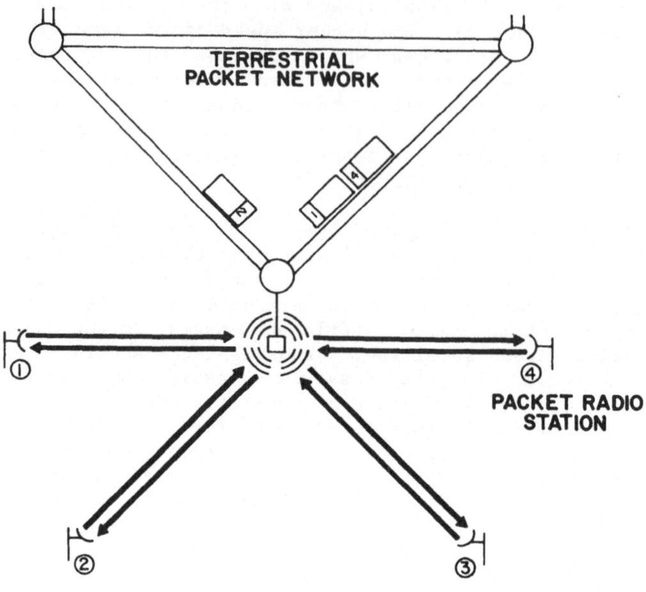

FIGURE 5

well-known tecnique to achieve high utilization of the available
transmission bandwidth is called CSMA, or Carrier Sense Multiple
Access. With CSMA a packet radio station will only transmit while it
is not observing the transmission of some other paket radio station.
Since the packet radio stations are normally using directional antennas
to save power and thereby cost, the central station has to send a busy
tone on a separate frequency to inform all packet radio stations of the
fact that one of them is transmitting. Collisions due to propagation
delays and transmission errors are recovered from by retransmission
after a random timeout interval.

LOCAL AREA BUS SYSTEMS

Local area bus sytems (LABS) represent another area where packet
techniques can be employed successfully. The LABS consist of a
standard coaxial cable with taps for the connection of different types
of devices. All devices share the transmission medium in a way that is
very similar to the packet radio system. Information is assembled into
packets and transmitted when the coax cable is idle. When packets
collide due to propagation delays, the cable retransmit mechanisms are
employed to recover.

The LABS is well-suited for the transmission of packetized voice.
Since any station can with equal ease transmit to any other station on
the bus the LABS act not only as a transmission medium but also as a
switching device. Standard interfaces for voice (T1 carrier PCM
format) and data (X25) can connect the LABS to other networks.

TELEMAIL

Electronic mail service has evolved as an exciting application of the packet technology. GTE Telenet recently introduced an electronic mail service called Telemail, which uses the public packet-switching network for the collection and distribution of electronic mail.

Managers effectiveness is limited by the effectiveness of the communications tools at their disposal. For example, a great deal of valuable time is consumed by the telephone. Incoming calls interrupt their work, often unnecessarily. Outgoing calls are unsuccessful 75 percent of the time, simply because the other person is unavailable . Multiple time zones worsen the problem by reducing the amount of time during the day when they may reach people in other parts of the country.

Telemail is designed to be used either by the manager himself or by the manager in conjunction with his administrative staff. Users talk to Telemail using everyday office terminology such as SCAN, READ, SEND, and FILE. Anyone can master the basics in less than 30 minutes of instruction and practice. Two levels of 'prompting', novice and expert, enable more experienced users to take advantage of many time saving shortcuts.

Any type of printed information can be sent via Telemail, including not only memos, but a wide variety of special reports and forms. When reports or forms are involved, Telemail prompts the user for the specific information your company requires, and even does some preliminary checking to assure that the answers make sense.

Telemail provides unlimited message storage for any period of time. Each individual subscriber can set up his own set of files or use a common department or company-wide electronic filing system within Telemail. After he sends or reads a message, he may store it under as many as six different categories of his choice. Storage costs only pennies per month for each message.

Messages are delivered to the user's electronic mailbox to be called for and read at his convenience in his office, his home, or wherever he may be. As an option, messages can be delivered in printed form directly to a specific data terminal, telex or TWX machine or by Mailgram.

Each user selects a personal password which may be changed instantly at any time. Unless the user tells his password to someone else, only he can access his messages.

Group lists can be set up as a shortcut to addressing each recipient by name. For example, lists can be set up for specific departments, locations, project teams, or for all employees as a group.

Telemail provides an electronic version of the telephone directory to help users locate other subscribers inside and outside your organization. A user can request to remain unlisted or to identify his user name only to those within his own organization.

Wide-Band Digital Communication for the 80's

A. Zeitz
Arlington, USA

Summary

Telecommunications needs have changed dramatically both in West Germany and the United States over the last twenty years. Although the West Germany Bundespost has a detailed study and plan to meet coming telecommunications needs, the United States must rely on private enterprise to answer the problem. This problem is further compounded by the size and decentralized geographic distribution of users in the United States.

One of the solutions proposed is the Xerox Telecommunications Network (XTEN), a nationwide high speed digital network accommodating data communications, document distribution, and teleconferencing transmissions. The system uses a combination of satellite or terrestrial facilities for long distance transmission coupled with microwave radio for local distribution.

Components of the network include the subscriber access unit, local microwave, transceiver, city or switching nodes, earth stations, a network control center and of course a satellite. Subscriber terminals are owned by the user but do not have to be identical to use the network due to its ability to match dissimilar speeds, codes and protocols.

The network control center in addition to its generic functions handles routine customer billing and accounting functions as well as changes or requests for service. The city nodes can act as central switching centers but may also store or process messages. The subscriber access unit or equipment interface is where the necessary speed and code conversions are performed so that transmission compatibility is effected between dissimilar terminals.

Another unique feature of the network will be the frequency reuse plan known as cellular radio. In dense metropolitan areas, each city node would broadcast in from different frequency quadrants in a six mile radius. Subscriber stations would communicate with the node through directional antennas.

There are economic and technical risks however. The 10 GHz bandwidth, used for the local microwave distribution is presently being used for other purposes and permission must be granted by the FCC for reallocation. The bandwidth has never been used for this purpose and there are many unknown technical problems that must be solved. Lastly, all of the financial unknowns must become knowns because Xerox is responsible to its stockholders.

To answer the problem of internal (intra-building) networks Xerox recently introduced ETHERNET, a passive loop with access devices that enable terminals to get into the loop. Further, Xerox, Digital Equipment Corporation, and Intel Corporation are cooperating in the joint development of a standardized, non-proprietary controller for ETHERNET which should result in considerable savings for the end user.

As for applications, two examples would be newspaper publishers who would utilize the network for remote printing of their products, and manufacturers with their facilities located over a wide geographical area, having diverse transactions requiring multiple location involvement.

Zusammenfassung

Auf dem Gebiet der Telekommunikation hat es im Verlauf der letzten
zwanzig Jahre dramatische Änderungen der Benutzeranforderungen gegeben. Dies gilt für die Bundesrepublik ebenso wie für die Vereinigten Staaten. Während in der BRD die Bundespost als zentrales Organ
detaillierte Studien vorgenommen hat, um den zukünftigen Telekommu-
nikationsanforderungen gerecht werden zu können, sind die USA auf
die Initiativen privater Unternehmungen angewiesen, die sich dieser
Problematik stellen. Die Problematik in den USA wird verschärft
durch eine größere Anzahl zu erwartender Benutzer und durch deren
weite räumliche Verteilung.

Eine der vorgeschlagenen Lösungen zu diesem Fragenkomplex stellt
das XEROX TELECOMMUNICATIONS NETWORK (XTEN) dar, ein landesweites
digitales Netz hoher Übertragungsrate, welches für Datenkommunikation, Dokumentverteilung und "Telekonferenzen" vorgesehen ist.
Für die Übertragung über weite Strecken werden Satelliten und entsprechende Bodenstationen eingesetzt. Für die lokale Datenvertei-
lung wird die Übertragung mittels Mikrowellen durchgeführt.

Die Netzkomponenten umfassen Netzzugangseinheiten für den Teilnehmer, lokale Radiostationen, Sender/Empfängerstationen, Vermittlungs-
knoten in den Städten, Bodenstationen, ein Netzsteuer- und Kontrollzentrum und natürlich einen Satelliten. Die Teilnehmer-Endgeräte
sind Benutzereigentum. Es kann sich dabei um unterschiedliche Gerä-
tetypen handeln, dank der Anpassungsmöglichkeit des Netzes an verschiedene Geschwindigkeiten, Codes und Protokolle.

Außer seiner eigentlichen Aufgabe bearbeitet das Netzsteuer- und
Kontrollzentrum die routinemäßige Gebührenerfassung und -abrechnung
sowie Änderungen und Nachfragen nach Dienstleistungen. Die Stadtknoten nehmen nicht nur die Rolle einer zentralen Vermittlungsstelle
wahr, sie können auch Nachrichten speichern und verarbeiten. Die
Netzzugangseinheit des Teilnehmers ist in der Lage, Geschwindigkeits-
anpassungen und Code-Konvertierungen vorzunehmen, so daß eine Über-
tragungskompatibilität auch zwischen verschiedenen Endgeräten be-
wirkt wird.

Ein weiteres herausragendes Merkmal des Netzes besteht in der Wieder-
verwendung von Frequenzen, eine Vorgehensweise, die unter dem Begriff "cellular radio" bekannt ist. In Großstadtbereichen sind Aus-
strahlungsbereiche von 6 Meilen Radius für jeden Stadtknoten vorgesehen. Entsprechend sind die verschiedenen Frequenzen zu vergeben.
Teilnehmerstationen kommunizieren mit ihrem Stadtknoten über Richt-
antennen.

Jedoch gibt es noch wirtschaftliche und technische Risiken. Die
10 GHz-Bandbreite, vorgesehen für die lokalen Funkstationen, wird
z.Zt. für andere Zwecke benutzt und eine entsprechende Umwidmung
bedarf der Zustimmung des FCC. Diese Bandbreite wurde noch nie für
den geplanten Anwendungsbereich genutzt, so daß noch eine Reihe un-
geklärter technischer Probleme zu lösen sind. Ferner müssen noch
viele finanzielle Fragen geklärt werden. Xerox muß seinen Aktionären
dafür geradestehen.

Außer dem globalen gibt es noch den lokalen Netzaspekt, z.B. die Verteilung der Information innerhalb von Gebäudekomplexen. Hierzu hat
XEROX kürzlich seine Lösung vorgestellt in Form von ETHERNET. Das
ist ein System bestehend aus einer passiven Schleife mit Zugangsein-
heiten, die es gestattet, Datenendgeräte an diese Schleife anzuschlies-
sen. Ferner entwickeln die Firmen XEROX, DEC und INTEL zur Zeit gemeinsam eine standardisierte allgemein erhältliche Steuereinheit für
ETHERNET, die zu erheblichen Einsparungen auf seiten der Endbenutzer
führen sollte.

Als Beispiele für Anwendungen sind Zeitungsverleger zu nennen, die
das Netz für das abgesetzte Drucken ihrer Erzeugnisse verwenden können,
und Unternehmen, deren Niederlassungen über größere räumliche Bereiche
verteilt sind und die Transaktionsbedarf zwischen diesen Bereichen
haben.

<u>Introduction</u>

Twenty years ago there were less than 6 million telephones in the
Federal Republic of Germany. Today there are almost four times
that many. And twenty years ago the need for telex, and advanced
high speed telecommunication systems in general were not nearly
in demand as they are today and will be in the 1990's. Twenty
years ago things in the United States were different. Cars had
tailfins on them and weighed over two tons apiece.

The transistor, although already about ten years old, was only
just beginning to find limited commercial use. And the country's
telecommunications system, largely confined to voice transmission
and telex was keeping pace with the needs of its users, for computers, high speed terminals and facsimile were in their infancy -
certainly not ready or able to communicate with each other.

Both the United States and the Federal Republic of Germany are
addressing the same problem then - how to be prepared to handle
huge amounts of information and how to get that information where
it needs to be - when it needs to be. I have briefly read the
telecommunications report published by the Federal Ministry of
Posts and Telecommunications in 1976, so I am somewhat familiar
with what the Bundespost is doing about the problem.

Because United States communications systems are privately owned,
however, there is no such overall plan although there have been
plenty of individual solutions and advanced communications systems
proposed to solve the problem of information movement and handling.
And because these concepts are advanced by individual companies,
they must assume the economic risk of their ultimate success or
failure.

We also have another problem in the United States. Fifty or even
30 years ago businesses were generally centralized in one geographic
location. Headquarters and manufacturing and sometimes even
distribution were all located within a short distance of each
other, making communications both written and oral, fairly easy.
Today American businesses are greatly decentralized and have
become multi-national. Manufacturing is where labor and raw

materials are available at low cost. Distribution is as close
to major markets as practicable. Data Centers (new in the last
twenty years) are located regionally to record, bill, keep inventory and do all the things necessary to keep a business
running smoothly, and in the process generate mountains of
information.

The Xerox Telecommunications Network: An Overview

One of the systems proposed to manage and handle the above
mentioned huge information flow is XTEN - an acronym for the
Xerox Telecommunications Network.

On November 16, 1978 the Xerox Corporation filed a Petition with
the United States Government's Federal Communications Commission
to request a reallocation of the little used 10 GHz microwave
radio band for use as an electronic message system. It had become
apparent to Xerox and many others that the U.S. telecommunications
voice network, never designed to carry vast amounts of digital
information, was becoming overwhelmed.

Basically the XTEN concept is a nationwide high speed digital
communications network accommodating data communications, document
distribution and teleconferencing transmissions. The network would
use a combination of satellite transmission facilities or terrestrial facilities and microwave radio for local distribution. We'll
be leasing satellite capacity from appropriate vendors, bringing
broadband service of up to 1,5 MBits to the user's site.

Figure 1 shows the XTEN communications link in its entirety.
Transmissions emanate from and are received from subscriber
locations both in the foreground and background cities. They
leave the subscriber location (1) on the 10 GHz microwave portion
via a small rooftop dish, two feet in diameter. In the foreground
city you can see one building (2) that is higher then the rest.
This is the city node - the switching center that makes the
determination where the message goes. In this diagram you can see
signals going between buildings and out to the earth station (3)
for long distance transmission. The earth station then transmits
to the satellite (4) which relays the signal to a distant city.

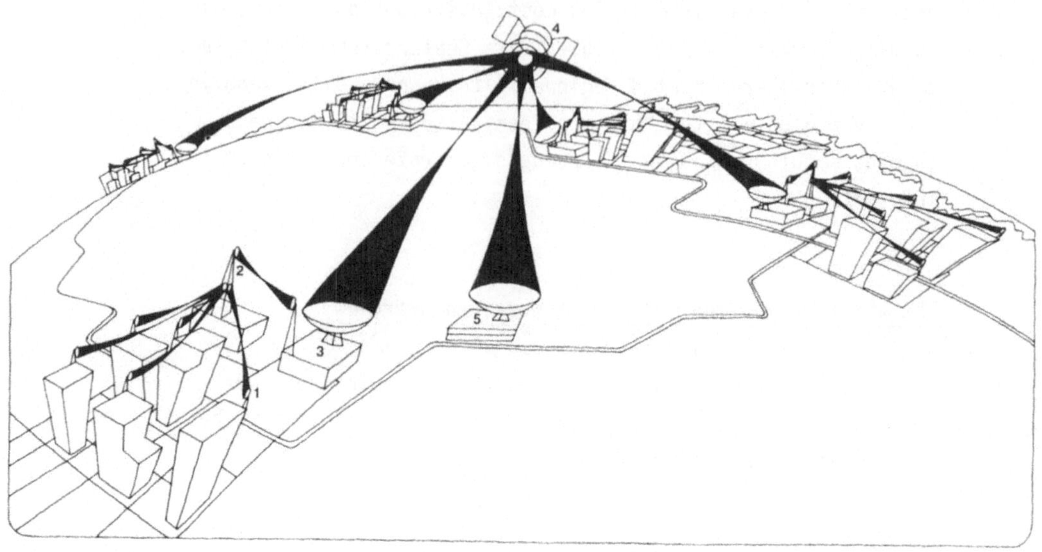

Fig. 1

CITY-TO-CITY - In connection with a petition it has filed with the Federal Communications Commission, Xerox Corporation has announced plans to establish a high-speed, nationwide digital information network. Customers will be able to send and receive large volumes of digital information between many locations at low cost. The Xerox Telecommunications Network (XTEN) will employ leased satellite capacity, radio links and communications processing capability. Typically, a customer's message will move from his terminal through Xerox-supplied equipment interfaces on his premises to transceivers linked to a rooftop antenna (1). From there, the message will be beamed to a sub station or city station (2) and then to an earth station (3) for transmission to a satellite (4). At the destination site, the message will travel a reverse path to the receiving terminal. At the customer's option, documents, messages and data will also be transmitted to a network control center (5) where they will be recorded for subsequent retrieval.

To the right of the earth station you see another earth station (5) communicating with the satellite. This is the network control center monitoring the performance of the network and providing a centralized resource for customer-oriented information such as billing, accounting, and system management practices to assure that customer-originated reports for service or equipment changes can be accommodated in the network.

XTEN Network Components

Let's take another look at the major components associated with the XTEN network (see Figure 2). On the lower left is the subscriber

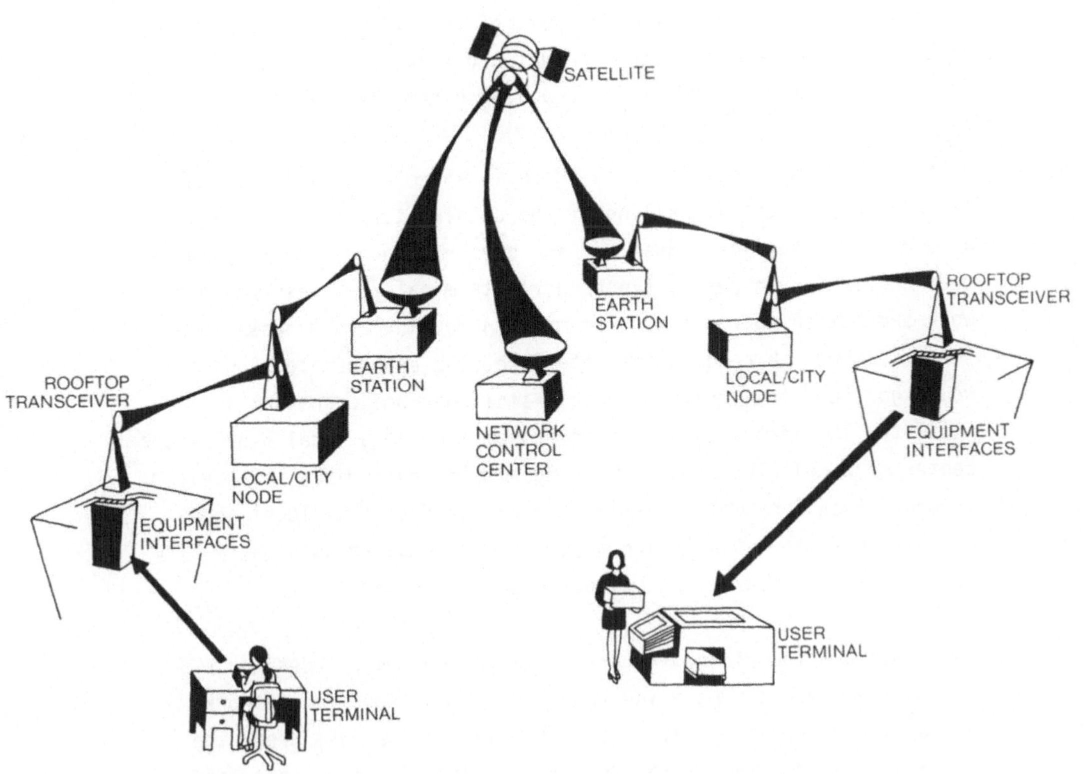

Fig. 2

RADIO LINKS - In connection with a petition it has filed with the Federal Communications Commission, Xerox Corporation has announced plans to establish a high-speed, nationwide digital information network. Customers will be able to send and receive large volumes of digital information between many locations at low cost. The Xerox Telecommunications Network (XTEN) will employ leased satellite capacity, radio links and communications processing capability. Typically, a customer's message will move from his terminal through Xerox-supplied equipment interfaces on his premises to transceivers linked to a rooftop antenna. From there, the message will be beamed to a sub station or city station and then to an earth station for transmission to a satellite. At the destination site, the message will travel a reverse path to the receiving terminal. At the customer's option, documents, messages and data will also be transmitted to a network control center where they will be recorded for subsequent retrieval.

terminal - in this case, a communicating word processor. With the XTEN concept, this terminal is connected directly to the equipment interface unit at the customer's facility. Here is the first point of entry into the XTEN system. I might point out that this diagram shows, on the lower right, the receiving terminal which, in this case, is depicted as a communicating document printer or processor. As you can see, each terminal is dissimilar. Because the XTEN concept proposes intelligence as part of the network, it could provide the services to match these two dissimilar terminals so that from a user's standpoint, they are compatible.

Back to the equipment interface. Here is where network intelligence would interpret the signalling of the sending terminal, identify the nature of the service requested, and perform the necessary speed and code conversion to permit the sender's terminal to access the network and obtain appropriate transmission routing and processing services. From this equipment interface, the appropriate signals would go to a transceiver and to a small parabolic antenna in the facility. The information is then transmitted to a local node at the center of a cellular area containing similar subscriber transceivers in other locations within a six-mile radius from this local node. At this node, the signals are repeated and combined into intra-city nodal links which are then collected at a city node.

Here possible functions of storage, switching, or processing, depending on the service being required and the specific network design would be performed. In our example: in the case where the message is to be transmitted to a recipient in some distant city, the city node prepares this information for transmission over either a satellite link, as depicted here, or alternatively, over highspeed terrestrial communication links which terminate at another city node.

In the center of Figure 2 is illustrated the Network Control Center. We have already discussed its function.

The right-hand side of Figure 2 is basically a mirror image of the left. Representing the components of the receiving end of this communication - in this case, an earth station is depicted with its city node, relaying the information on through a local node to a rooftop antenna at the destination, and from there, on to the equipment interface and ultimately to the printing unit for which the message is intended. By this point, any necessary transformation services could have been provided on the data so that it is now compatible with the equipment interface at the user site. Additionally, if the receiving unit had been busy, the system would have stored the message, and, based on the grade of service requested by the sender, would have either attempted delivery at a later time, at an alternative location, or awaited further instructions from the sender.

The Cellular Radio Concept

The uniqueness of the proposed XTEN concept and the key enabling
capability for the provision of high-quality, wideband, digital
data services, is the use of microwave radio for local distribution
of information. The present XTEN design employs a "cellular"
configuration of omnidirectional transmitting and receiving local
nodes located throughout a metropolitan area to provide direct,
two-way communications with transceivers located at subscribers'
premises. As depicted, the local node stations would be linked to
a city node using conventional narrow-beam microwave radio circuits
for the intra-city communications links between the local node and
city node - it is to provide for these two segments - that Xerox
has requested the reallocation of spectrum.

The local distribution system thus uses cellular radio to allow
reuse of radio frequencies within a given geographical area and to
assure complete city coverage. In a variety of metropolitan areas,
XTEN, and systems similar to it, will use more than one local node.
The resultant reuse of frequencies from node to node will be important in terms of overall spectrum management in the use of this
very scarce national resource. Depending on the planned traffic
density and pattern of traffic flow, interference problems could
be mitigated by judicious arrangements of the stations and by
frequency assignment patterns which have been proposed within the
Xerox Petition to the FCC.

To illustrate how this works, the Boston, Massachusetts metropolitan
area would have a possible coverage pattern by XTEN local nodes
using the 10 GHz spectrum. Each circle would represent the broadcast
pattern of an individual local node. Distance covered by each pattern
is a radius of approximately six miles, limited by both the technical
factors associated with power, and attenuation factors which tend to
decrease the margin or signal-to-noise ratios for RF transmission in
the presence of various atmospheric conditions.

Subscriber stations would have directional antennas aimed toward
the appropriate local node. Every subscriber station in a sector
will receive all transmissions from the node to that sector.
However, each subscriber station will accept and decode only
those transmission bursts addressed to it. When a subscriber

station signals to its node that is has information to transmit,
the station will be assigned a time slot and will transmit its
information in bursts at a minimum of 256 kilobits per second.

This is a very efficient way to utilize microwave radio, and
provides multiple area coverage with the same frequence, rather
than requiring a frequence allocation of five or six times the
amount of bandwidth as might otherwise be required to cover a
large metropolitan area. The addition of local nodes to a city,
and the proposed transmission patterns would permit future
service demands to be met.

Because this idea is unique and innovative it also carries a
certain amount of risk with it. Since filing our petition over
a year ago we have learned a lot more about the economic and
technical problems associated with the proposed XTEN system,
especially in the area of microwave radio. Some of those problems
appear quite formidable and could require a re-thinking of the
entire network design.

The fact is that the 10 GHz band has never been used for an
electronic message system as we propose, and there are a lot
of unanswered problems as far as testing and feasibility. Will
we have problems for example with hail storms or heavy rain
interfering with the system? We have some theoretical answers
to those questions, but until actual tests are carried out,
they remain as unknowns.

Since we are responsible to the owners of our company - our
shareholders - as prudent businessmen and women we must ascertain,
as closely as possible, what our finacial exposure will be. We may
decide to go with the original concept or modify it or get into
a network of a different sort.

Intra-Building Communication

What about communications within a building? How do they inter-
connect with XTEN? Is there a reduction in speed or quality
internally or when connecting to the XTEN network?

Figure 3 shows a diagram of the various subscriber terminals
in a building and how they would be connected to the XTEN
subscriber access unit and rooftop antenna. There's the CPU,
some communicating word processing machines, the mailroom with
an intelligent copier, a bank of data terminals and on the
right side the teleconferencing facility in use.

What's interesting about this diagram is the fact that all these
devices are connected to a single wire or link which in turn is
connected to XTEN. There has been a great deal of concern about
how incompatible devices will be able to talk to each other within
a given location be it a single floor of a building or an entire
building "campus".

As one solution, Xerox recently announced a product called ETHERNET,
which is basically a passive loop with an access device that enables
a terminal to get into the loop. The loop can be up to 1500 meters
in length. Speed is up to 10 megabits.

Furthermore, just recently Digital Equipment Corporation (DEC),
Intel Corporation and Xerox announced plans wherein DEC and Intel
could develop, on a non-proprietary basis, the necessary transmission
and controller device for operating with ETHERNET. This will result
in a considerable savings as well as the establishment of a de-facto
industry standard for intra-building communication networks.

Some Examples of XTEN Applications

How will American Companies be able to take advantage of this
unique high speed network? Let's take a look at a couple of
American businesses and see how they would apply XTEN to their
communications problem.

A number of newspaper publishers in the United States are applying
high speed satellite communications links to the production of
their newspapers. The Wall Street Journal, for instance, transmits
full editions from Carlstadt, New Jersey simultaneously to Palo
Alto, California and Chicopee, Massachusetts over satellite links.
These locations represent satellite printing plants of the Journal.

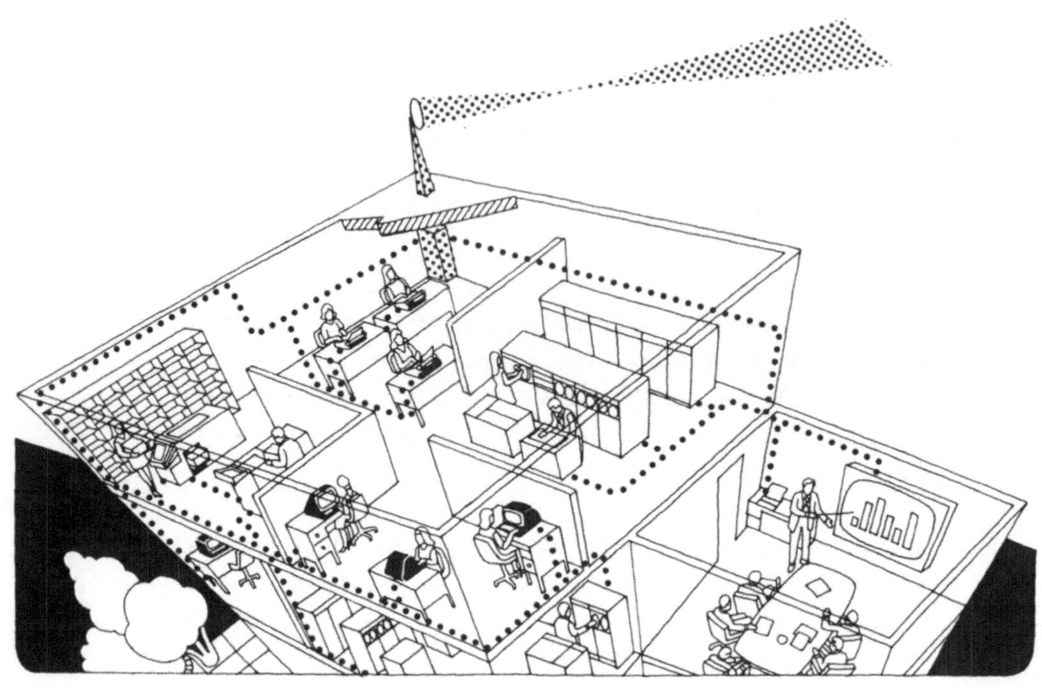

Fig. 3

WELL CONNECTED - Three principal applications of the proposed Xerox Telecommunications Network (XTEN)
are illustrated in this cutaway of a typical customer facility. <u>Document Distribution</u>: A wide variety
of sending and receiving terminals and work stations can be linked because of the network's
compatibility features. Advanced printing devices at mail distribution centers (left) allow the rapid
delivery of documents to multiple recipients. <u>Data Transmission</u>: From interactive terminals (lower
center), data can be sent and received with private-line quality on request basis. A variety of remote
job entry terminals can utilize broadband service to permit high-speed input/output operations.
Equally high-speed computer-to-computer applications (top center) allow processors and files to be
located where they best support a customer's business. <u>Teleconferencing</u>: A working conference
atmosphere (right) will include high-quality, still-frame video, hard copies on demand and voice
capability. All data is routed through an equipment interface to a rooftop antenna.

They are expected to put more of these plants into service in the
future. Denver, Colorado and Dallas, Texas are under consideration.

At the recent American Newspaper Publisher's Association/Research
Institute convention in Atlanta, Georgia it became known that many
of the larger chains have plans or are actually implementing the
establishment of satellite printing plants to lower costs and increase
efficiency. With the establishment of a nationwide digital high speed
communications network especially one with high speed local

distribution capabilities like XTEN, many more would be able to cut their printing costs.

Let us consider how a satellite printing plant located outside a city might be linked with the downtown editorial and makeup offices located in the background. This link could be up to 1,5 megabits wide - if you needed that capacity - and could be realized as a direct point-to-point microwave link.

But let's suppose your regional printing plant is located hundreds or perhaps thousands of miles from headquarters. In an overall view of the XTEN model (as shown in Figure 1) the publisher's transmission would travel from the city in the background over to the earth station on the upper right center of the photo - up to the satellite at the top of the picture, and in reverse to the earth station - left center and out to the plant via microwave link located to the left, all accomplished without costly investment in large earth station facilities.

As a second application let's see how a U.S. truck manufacturer might use XTEN in running his business.

Here's the classic example of a decentralized business - manu-facturing in one location, distribution located regionally - as close to the major markets as possible, regional data centers supporting the sales and service organization located in major market centers, and headquarters - located close to the financial community in a large city. With XTEN as a primary communication link between these various locations this company could process orders, take inventory, maintain service and parts operations, control budgets and a host of other functions speeding up these transactions while lowering their overall communications costs.

For instance, to process an order, a high volume dealer located in a small midwestern city, could place his customer's order through XTEN directly with the factory. Using his own terminal hooked into the network through the XTEN owned SAU centrally located in the city, the information would be relayed by the city node to the earth station located outside the city. From there, the earth station relays it to the satellite which in

turn sends the information to the earth station on the receiving end. The order then proceeds directly to the factory. There are other departments within the company that can simultaneously receive copies of that order.
- The regional distribution center to see if the make and model is available from stock.
- Headquarters, for the daily and monthly sales and marketing reports as well as financial reports.
- The regional data center for inventory control and regional sales records as well as sales commissions.

Thus with the flexible addressing options which will be offered with the network services, all interested departments within the company would know overnight or sooner that an order had been placed.

What you've seen here today - the conceptual description of the Xerox XTEN domestic network and how it might be used - represents a proposed solution to the mounting problem of managing information in the United States. Since filing the Xerox Petition nearly a year and a half ago, we have made good progress with the FCC. We anticipate a ruling on the reallocation some time before the end of this year. We are testing the radio transmission system in a number of U.S. cities. Frequency clearances and site selection are well underway in major U.S. metropolitan areas. We look forward to helping businesses become more efficient and productive by having a better way to manage information.

<u>DISCUSSION after the lecture of Ronald A. Zeitz</u>

Question: I have two questions. You talked about spectral management
and I am sure the FCC can assign to you only a limited
number of frequences at 10 GHz. First question is: How
many frequences are there. Second question is: If the
number of frequences is not sufficient or if you want to
distribute from one of the final nodes to a number of
subscribers' premises or to a large university campus you
need cable. Are you allowed to lay these cables of whatsoever
kind crossing street etc. by yourself or have you to rent
these lines from the local telephone companies?

Answer: First of all let me explain that a lot of that I talked
about and which I showed you is extremely dependent of what
the FCC decides to allocate. We have applied for the 10.55
to 10.68 GHz band. Within that we have applied for ten channels
of 120 MHz. The very surprising comments that come back from
other interested parties, manufacturers, users and
governmental groups show us that we would only need 60.
We then have readjusted our requirements so that we can now
handle 90. Some of the bandwidth is needed for overhead,
for the management of the network and for keeping it
efficient. We try to make that point clear to the FCC. So
we really need 90.
As far as the cable problem goes I don't think I have a
sufficient answer for that. We talked about the ETHERNET
as being a possible solution for connections. Well, ETHERNET
is another XEROX product, it is not part of XTEN. I want to
make this quite clear. It is being marketed, sold and
engineered by a different group within the company.
However, we present it today as a possible solution to this
problem of intra-building and intra-campus communication.
But how that is being marketed and installed, I really
don't know.

Question: You concentrated in your talk on transmission and data
distribution problems. But there are a lot of computers
already around and applications on top of that computers
and existing terminals and there is a hell of problems
to connect this existing equipment to new networks. So,

what kind of adaption, what kind of philosophy do you
have to connect what is already there to your new
service? And may I add a question to Mr. Opderbeck -
I don't know whether he is able to answer it - I would
like to hear what percentage in software development
effort has been spent into network software on one hand
and into adaptation software on the other hand.

Answer: I'll try to answer that also. During the course of the
presentation I mentioned that this (=XTEN) is an intelligent
network. We are designing it, of course, with what is available now around today. Everybody minds that because it is an
immediate problem. But also - what we feel - we must face
the problems of tommorow. We will be able to handle the
already existing systems as well as the equipment people
are just talking about now. But that will involve a period
of time. Now to your second question. I would say: an
enormous amount. A rough estimation would be 1/3 for the
switching and network management software and about 2/3 for
the adaptation. That high portion being due to missing
standards.

Question: What do you expect from the utilization of own transmission
facilities instead of the BELL company or other carriers?
What does this bring roughly for the customer of your
system. How can the huge investment pay off?

Answer: Let me answer the second question first. I think that both
are going into the same direction. Telenet is evolving,
Tymnet has evolved and XTEN has not evolved yet but will
evolve and will add services as the time goes along. It
will take some time until we have competitive things to
offer you. The system XTEN will provide earth stations,
subscriber channels, subscriber access units,
the microwave transmitters and receivers, the network
control center. That is an enormous amount of capital
investment. And we are trying to offer a low cost
service by offering then low capital investment for
the users all in terms of low cost terminals, computers
or what there will be in the future. These are basically
economic considerations. Secondly, if we are the custodians

of the network or the owners of the network we are probably
able to offer a higher degree of cellular services and
more reliability because we then have the supervision and
the direct control. There has been a lot of investigations
inside and outside the company. Much of the amount of the
capital investment depends on the still expected decisions
of the FCC and there has been some speculation. What we
are doing right now is that we try to evaluate the time
phaisng of this network so that we get not a too huge
economic burden. So there are a lot of managerial problems
besides the technological ones that are yet unsolved
although they are workable. All these things must fit to
the FCC proposal. But there will be a point and after that
point (- and I don't know what that point is and I am not
sure whether anybody in the corporation knows -) there will
be a break-even-time and we will start to make money with
the network.

Question: What is the real "added value" of your system? You spoke
a lot about transmission but nearly nothing about the
services and the facilities you offer the customers. What
are the new added values of your system compared with e.g.
the existing TELENET or TYMNET?

Answer: We are in the process of evolution with XTEN. There is
another slide which lists a whole group of services. We
found the listing of services has come in conflict with
some investigation the FCC has done. What the FCC tells
us covers the whole bandwidth how the welcome look like
and how much we can have und how much we should have.
Those are nice factors which are not yet under control
at this time. As I said before if the FCC will have made
its decision we all are waiting for we will be in a much
better position to define what the services will be
precisely we are going to offer. For instance I mentioned
multiple addressing as one of our services and I showed
an example of that, store-and forward services, three
types of delivery (immediate, few-hours-delivery,
organized delivery). These are the types of things we
are getting into.

Question: If you have a kind of broadcasting system where each
receiver is picking out of the whole set of messages
that part which is determined for them, how do you ensure
from the legal standpoint data security and confidentiality
of the information?

Answer: In the U.S. that question has come up in many presentations
to potential customers and users, and the answer is not an
easy one. If we would make a list of the ten most desired
services or most desired capabilities of the network,
security would not be number one except by certain
government agencies who insist on it. Number one happens
to be reliability. Security, as far as encryption goes
will be a service that we are going to offer, but this is
not a critical question at this time since this is not one
of the highly desired services.

Question: What is the data rate you provide for teleconferencing?

Answer: The transmission rate will be 256 kb/s. We will provide
a freeze frame mode without motion with probably a
2 or 3 per second refresh rate on the screen. This we
have found to be more than adequate for most presentations
with graphic material. We also found that most people
don't realize that motion is not needed for business
purposes not even space. We have decided to go with the
two-screen mode.

Question: What about speech? I didn't notice any telephone in your
pictures.

Answer: Voice has not been part of the network designed. We find
that the existing telephone-network as such is adequate
for voice communication today but not adequate for data-
communication.

Satellite Integrated Communications Network

J. J. Duby
Paris/France

Summary

The topic of the presentation is "Satellite Integrated Communication
Networks". Each of the four words is important: we will consider
communications using geostationary satellite relays to transmit all-
together voice, data, image, and text in an integrated way over a
multi station network. One of the most publicised examples of such
networks has been SBS (Satellite Business Systems Corporation) that
is about to be put into service in the United States, but other
similar networks are being planned all over the world; (e.g. Telecom I,
ECS, etc. ...); they will be mentioned in the course of the presentation.

The purpose of the presentation is simply to show the basic concepts, to
give an understanding of the main technical problems that face satellite
integrated communication network designers, of the different approaches
that were followed to solve them, and of the main economical implications
of such networks. In the first part we shall focus on the main technical
problems (e.g. transmission, frequency, power, multiplexing techniques,
space channel allocation), the present solutions, the future trends. - In
the second part we shall deal with the economics, adding some emphasis
on "value added" services and network management. Throughout the
presentation, we will use the SBS system as an example to demonstrate
and illustrate some of the notions and techniques described.

Zusammenfassung

Anhand des Beispiels SBS (Satellite Business Systems Corporation) wird
ein Satelliten-integriertes Kommunikationsnetz vorgestellt, in dem über
einen geostationären Satelliten zugleich Sprach-, Daten-, Bild- und Textkommunikation erfolgen kann.

Der Beitrag versucht, die grundlegenden Konzepte solcher Netze darzule-
gen. Im ersten Teil wird auf die wichtigsten technischen Probleme einge-

gangen. Dazu gehören die Technik der Übertragung, der verwendete Frequenzbereich, Verstärker, Antennen, Multiplextechniken und Kanalzuteilungsstrategien. Der zweite Teil des Vortrags streift wirtschaftliche Aspekte wie z.B. die angestrebte Gebührenstruktur.

1. Technical approaches

To implement a satellite integrated communication network, one has to
solve three basic technical problems: how to transmit to and back from
the satellite; how to multiplex into the space channel (and demultiplex
out of the space channel) such different input devices as telephones,
computers, copiers, printers, keyboards, display terminals, etc. ...;
how to manage the space channel to optimize its use among all the users.
Let us cover those three areas.

1.1 Transmission

Basically, satellite communication is identical to microwave transmission. Because of the earth curvature, microwave transmission has to
use relay towers at regular intervals. Satellite transmission uses one
big relay tower 22,000 miles up in the sky, i.e. a geostationary
satellite ... Each earth station transmits up to the satellite which
retransmits back to the receiving stations.

This being said, when one considers the three areas of frequency,
power and antennas, satellite transmission has some specific requirements.

1.1.1 Frequency

Like microwave transmission, satellite transmission has to take into
account the use of the frequency band by other users like radio communication, radars, TV or ... microwave ovens. Besides, it also has to
consider upper atmosphere characteristics that microwave transmission
need not be concerned with, like molecular resonance. Without going
into any deeper technical detail, this should explain why frequencies
used for satellite communications are chosen among three classes:

$$6 - 4 \text{ GHz (Giga Hertz or } 10^9 \text{ Hertz)}$$
$$14 - 12 \text{ GHz}$$
$$30 - 20 \text{ GHz}$$

For a satellite in the first class, for instance, stations will transmit up to the satellite at a frequency in the neighbourhood of 6 GHz while the satellite will transmit at around 4 GHz. Similarly, in the 14 - 12 GHz class (30 - 20 GHz resp.) the up link frequency will be near 14 GHz (30 GHz resp.). Note that the downlink frequency is always the lower one: this is because lower frequencies are less sensitive to atmospheric disturbances like rain and thunderstorm, since the signal coming from the satellite is already weaker due to the limited on board power, it is always allocated the lower frequency.

For the same reason, the first communication satellites used the 6 - 4 GHz band - and a majority of those in US today still do. However, as technology improves, more systems are using the 14 - 12 GHZ band which provides two main advantages: more bandwidth and possibility to use smaller antennas. Indeed, all the satellite integrated communications networks being implemented today use the 14 - 12 GHz band, e.g. SBS, Telecom I, ECS, etc..

To prepare for the need for even more bandwidth, some experiments are being performed in Japan and in Italy in the 30 - 20 GHz band.

During the coming decade however, satellite integrated communications networks will use mostly the 14 - 12 GHz band.

1.1.2 Power

As we have just seen, satellite communication designers have tried to improve the bandwidth by using higher frequencies. Another factor to increase the bandwidth is to increase the transmission power.

While increasing power on the ground is easy, increasing it on the satellite is the real problem. The key limitation factor in the satellite is the "payload", i.e. the mass that will be put into orbit. Within this mass tradeoffs can be made, for instance having more power cells to increase power or more rocket fuel to increase life time (the satellite position has to be constantly monitored and, if necessary, corrected firing mini rockets.) But generally speaking, "weight is the enemy".

In present state of the art, satellite repeaters - also called transponders - use Travelling Wave Tube (TWT) amplifiers. Modern TWT amplifiers are rated at a few hundred Watts (100 to 200 is common, 500 at experimental stage). This power allows a bandwidth in tens of Mbit/s (in SBS for instance, each transponder has a bandwidth of 40 Mbit/s).

As satellite payload increases, and as technological advances in other areas (e.g. more efficient solar cells) allow to give a bigger share to on-board communication equipment, the total bandwidth of the satellite is increased by increasing the number of transponders. For instance, while Intelsat I and II had only two transponders, Intelsat IV had 12 and Intelsat V has 27.

However, increasing the number of transponders creates a multiplication of separate channels with no communication between them: in fact, each transponder handles a separate network. To obviate this, techniques have been designed like transponder hopping (by which an earth station can switch from one transponder to another) and on-board switching (by which a transponder receiver output on the satellite can be switched to input another transponder transmitter). The former will soon be in operational use by satellite integrated communication networks (e.g. Telecom I); the latter may be more experimental. In any case, those techniques will become absolute necessities as the number of trans-ponders increases.

1.1.3 Antennas

The last area we shall consider, antennas, is of course not independent from frequency and power. Let us distinguish between on-board antennas and earth antennas.

There are two basic design approaches for on-board antennas: extended coverage and multiple spot beam. Extended coverage antennas radiate their power over the whole territory that is serviced by the network, regardless of the distribution of the earth stations. For instance, the satellite antenna of SBS radiates over the whole continental United States. One obvious drawback of this approach is that it pours

power over square miles of desert sands, thus wasting costly energy.
That is why multiple spot beam antennas have been designed which
concentrate their power over "useful" areas, i.e. where the users
are. This approach was made possible by recent progresses in antenna
design. It requires a much more precise positioning of the satellite
(using more rocket fuel - always the same trade-off). It also has
less flexibility in the sense that it is highly dependent on the
geography and may prevent new users to get service in non radiated
areas.

Regarding earth station antennas, their size is determined by the
frequency used and the power radiated from the satellite. 6 - 4 GHz
transmission requires 20 meter antennas, which are bulky, expensive
and require extremely precise tracking and pointing. 14 - 12 GHz
transmission, however, requires only small size antennas (a few meter
diameter) which cost less than 100 K$ and can be placed on roof tops.
This in itself is one of the major reasons why satellite integrated
communications networks chose the 14 - 12 GHz band.

1.2 Multiplexing

Typically, the bandwidth of a satellite space channel is measured in tens or hundreds of Mbit/s. Typical inputs into this space channel are voice (tens of kbit/s), data (from a few kbit/s to Mbit/s), image (up to several Mbit/s). It is therefore necessary to multiplex these inputs into the space channel and demultiplex them out of the space channel.

We shall consider essentially one transponder channel, the different transponder channels being separated by frequency and/or orthogonal polarization. Two basic multiplexing techniques are used: Frequency Division Multiplex (FDM) and Time Division Multiplex (TDM).

1.2.1 Frequency Division Multiplex

This technique is used in the first communication satellites. It is well suited for analog transmission of voice: the space channel bandwidth is divided into 4 kHz channels. Each 4 kHz channel is used to transmit a 3.5 kHz voice signal, with 0.5 kHz guardband between each voice channel to avoid interference (crosstalk).

Advantages of FDM are its simplicity and its adequacy to telephone transmission. It still is the most commonly used one in communication satellites like the INTELSAT family. However, FDM uses the bandwidth in a somewhat inefficient way (the guardbands are lost to information transmission) and, moreover, it lacks the flexibility required by data transmission.

Why does data transmission require more flexibility than voice transmission: essentially because voice channels are readily interchangeable while data channel bandwidth requirements vary with the data rates. A voice grade channel can transmit low speed data (up to 10 kbit/s) as is done frequently using modems over the telephone public switched network. But when medium speeds (e.g. 128 or 256 kbit/s) or high speeds (1.024 Mbit/s and above) have to be transmitted, FDM requires that several adjacent voice channels be combined together to provide the necessary bandwidth. If the voice channels occupancy rate is high (and it usually is ...) one may have to wait until conversations are over, or switch conversations to other channels, which requires time and trouble.

The problem is identical for image transmission. Television broadcasts which are transmitted over FDM communication satellites require reservations to be made in advance for several MHz bandwidth. If the bandwidth cannot be made available, you may miss the beginning of the football match on Eurovision... .

This is why communications satellites which transmit a lot of data, like integrated communications satellites, use Time Division Multiplexing rather than Frequency Division Multiplexing.

1.2.2 Time Division Multiplexing

FDM principle is to allocate to each individual user some of the space channel bandwidth all the time. TDM principle is the opposite: each user is allocated all of the space channel for some of the time.

In other words, each TDM user gets allocated not a frequency slot as in FDM, but a time slot. If it has a lot to transmit, its time slot will be longer and conversely. User time slots are allocated in such a way that time slots for users transmitting from a given earth station are contiguous: thus, each earth station will transmit a burst of data and listen to the bursts coming from the other stations.

To control the station burst timing and duration, station bursts and time slots will be allocated within a fixed duration frame. The allocation pattern of a frame will be repeated periodically until it is changed due to a change in transmission requirement from one of the stations (see Fig. 1).

Fig. 1   TDM frame structure

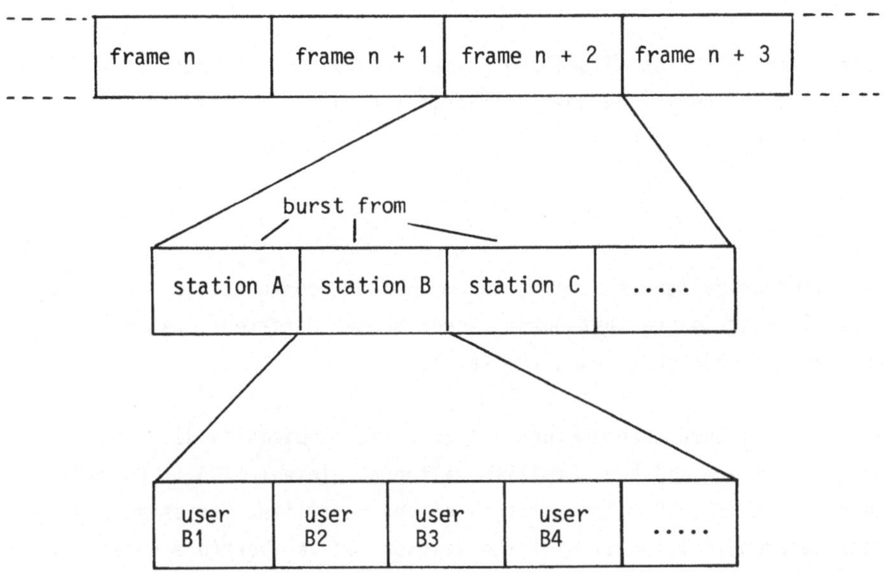

The duration of a frame is determined so that it ensures continuity of
voice and data transmission. The frame used by SBS for instance is 15 ms
(milliseconds). Thus, each station in the SBS system will buffer the
data (be it pure data or digitized voice or image) sent by its users for
15 ms and transmit it every 15th ms. Conversely, it will monitor the
bursts received from the other stations through the satellite, looking
for the user slots in each burst which are sent to its own users,
unbuffer these slots and transmit them over terrestrial lines to their
respective destination at the proper speed.

One readily sees that TDM is more complex than FDM. The reason why it
is nevertheless more widely used by integrated communication satellites
is that it allows a more efficient use of the bandwidth (the only guardbands required are between station bursts) - although one has to take
into account the necessary command and control information overhead -
and, more importantly, that it can react more quickly to changes in trans-

mission needs from the users - simply by changing the duration of their time slots.

We shall consider slot allocation in more detail in 1.3. Before, we shall focus on one of the most difficult technical problems of TDM: burst synchronization.

1.2.3 Burst synchronization in TDM

Burst synchronization is a key problem in a TDM communication satellite because it must insure that bursts, sent by two different stations, never overlap when they reach the satellite.

The problem is compounded by the fact that the propagation delay between each station and the satellite, although close to 125 ms corresponding to the 22,000 miles altitude of the satellite, varies according to the geographical location of the station. It is therefore necessary to precisely determine the propagation delay for each station.

Assuming that each propagation delay is precisely known, synchronization can be insured by a master station, which will conduct the orchestra of the other stations by sending at the beginning of each frame a frame reference burst. The frame reference bursts will indicate to the other station not only the timing of the frame period, but also where their respective bursts must be positioned within the frame and how long they may last. The so-called slave stations constantly monitor frame reference bursts from the master station to extract their own bursts timing information.

But, as said before, slave stations also need to know their propagation delay to insure proper burst synchronization. Two methods are used to get this information: open loop and closed loop synchronization methods.

In the open loop method, the master station determines the exact position of the satellite by triangulation, computes the distance between the satellite and each slave station, and finally transmits this information to each respective slave station.

In the closed loop method, each station computes its own propagation delay by listening to itself. More precisely, each station sends periodically a transmit reference burst which it listens to after it has been retransmitted by the satellite, then immediately deducts its propagation delay.

Open and closed loop methods have their own pros and cons. Open loop allows simpler (non intelligent) earth stations. It is also readily usable when an earth station cannot listen to itself as may be the case in spot beam coverage of satellite transmission. On the other hand, closed loop is more efficient as the information computed by each earth station is better up to date than through the master. It is also more reliable as a closed loop network can keep its synchronization for a limited period of time, while everything collapses in an open loop network as soon as the master station does. One can assume that the specialists will quarrel for a long time over which method is the best.

1.3 Space channel allocation

We shall consider space channel allocation only in the case of Time
Division Multiplex. In FDM indeed, either the problem is very simple
(allocating 4 kHz channels to telephone conversations) or, when more
complex, is handled through advanced reservations of large bandwidth.

In TDM, it is of course possible to have the same kind of static allocation: bursts are fixed within a frame or change only a few times per
day, based on day/night activity (e.g. West Coast/East Coast balance
in the USA) or on reservations for special needs. But the real advantage
of TDM is to be able to have dynamic allocation, i.e. to be able to
allocate bursts and slots within a frame according to the instantaneous
transmission needs of the users. This capability is more often referred
to as Demand Assignment (DA).

In TDMA/DA (Time Division Multiple Access/Demand Assignment) systems,
space channel allocation is done by the master station, based upon the
transmission requests which are received 250 ms after they are sent and
as any new allocation sent by the master station is received another
250 ms later, it is not possible to react instantly, e.g. from one frame
to another, to a transmission request sent by a station. It must be noted,
however, that a given station can reallocate its users slots within its
own bursts from one frame to another, at the only condition that it does
not overflow its burst boundary. This is the reason why frame duration
is chosen short enough to allow voice communication to be established
soon enough after a user starts or resumes talking so that the listener
does not notice that initial part of the speech was cut off - hence
SBS's 15 ms.

When a station needs a longer burst, however, as we have seen above, it
has to request it from the master. The presence of a transmission delay
which inhibits instant modification, has led to the notion of superframe.
To explain this notion, let us describe the SBS system superframe
structure.

An SBS superframe consists of 20 frames. It therefore lasts 300 ms, that
is a little longer than the average inter station propagation delay.
Burst allocations are made and are valid for one superframe - but, as
we have seen, slot allocation within a burst may change within a superframe. Changes from one superframe to another are handled by the
following mechanism:

- At superframe N, stations send requests to the master.
- At superframe N + 1, the master receives requests sent during
  superframe N and processes them, computing new allocations.
- During superframe N + 2, the master sends new allocations.
- During superframe N + 3, stations receive new allocations.
- At superframe N + 4, all stations transmit according to new
  allocations.

Thus, requests for new transmission needs are satisfied 1.2 seconds
after they have been sent.

To avoid this delay in most cases, and also to avoid unnecessary
increases/decreases of burst duration, burst times are allocated to
stations in increments corresponding to 224 kbit/s channel rate or
3360 bits length. As slots are allocated in 480 bit increments corresponding to a 32 kbit/s channel, a station always keeps a safety buffer
of unused burst space to allocate extra slots from where needed. It is
only if this safety buffer gets too small - or, conversely, too large -
that the station requests extra burst space - or releases burst space.

The most advanced integrated communications satellites use Time Division
Multiple Access with Demand Assignment, be it one level Demand Assignment at the master station only or two level Demand Assignment at the
master level and slave level.

Before closing this chapter, it may be interesting to describe the
third level in the space channel allocation method used by the SBS
system. Each SBS transponder is used by several networks, each network
corresponding to one enterprise user, the different stations in the
network corresponding to establishments of the enterprise. At the top

level, the space channel, i.e. the frame length, is distributed among
the different networks, leaving an unallocated pool - a safety buffer -
at the end of each frame (see Fig. 2). Each network can thus distribute
its allocation among its stations. In case extra capacity is needed on
a temporary basis, e.g. for a teleconference a high speed computer
look-up, it can be allocated out of the pool.

Fig. 2 SBS Frame Capacity Distribution

CUSTOMER				POOL
A	B	C	D	

2. Economic aspects of Integrated Communication Satellite Networks

In this second part we shall first describe the main economic characteristics of Integrated Communication Satellite Networks, then we shall
give a few details on SBS tariffs as they were filed to the Federal
Communications Commission.

2.1 Economic characteristics of ICSN

The main factor in the increasing success of ICSN's is the speed by
which their costs have been decreasing. To give just one example,
basic transmission costs have been decreasing at a rate of 25% per
year in the last 15 years: yearly cost of an INTELSAT I circuit was
30 K$ in 1965, 3 K$ in 1975 on INTELSAT IV A, down to 1 K$ in 1979
with INTELSAT V.

Independently from this trend, the economics of ICSN's have some
important characteristics when compared to classic terrestrial communications.

The first one is that satellite communication costs are not distance
sensitive. The break-even point above which terrestrial communications
become more expensive than satellite communication varies with the
different countries tariffs. To consider one example, SBS vs.
terrestrial communications in the United States break-even point is
at thousands of miles for voice and hundreds of miles for high speed
data. Generally speaking, the economic advantage of ICSN's does not
reside in communication cost savings. Main saving factors are first
integration, second networking. Integration allows bandwidth usage
levelling among the different transmission needs (voice, data image,
text) instead of having separate transmission media (voice lines, data
lines, image microwave, etc. ...). Networking ensures any to any
connectivity without complex wiring or store and forward; it also
allows load levelling among the different transmitting stations as we
have seen in 1.3. While the break-even point for basic transmission
is above hundreds or thousands of miles, the break-even point for an
ICSN with stations 500 miles apart is attained with three nodes only:
at three nodes and above, an ICSN is cheaper than a terrestrial network.

These savings can be further compounded with new applications like
teleconferencing (travel costs savings), data processing load levelling
(equipment savings), high speed fax (mail savings and work cycle
improvement), etc. ... . According to the different usages considered,
studies have shown that such value added capabilities could bring
savings several times greater than transmission cost savings.

2.2 A quick overview of SBS tariffs

SBS filed a tariff application to the FCC for three different services:

Communication Network Service A: this service provides intra enterprise
integrated communication networks with dedicated earth stations. In other
words,each earth station is situated on a customer premises (headquarter,
laboratory, plant, etc. ...) and used uniquely by this custumer.

Communication Network Service B: this service provides the same capability as the preceding one, with the difference that earth stations are
shared between different customers. CNS B is more accessible to users
whose communication needs are not large enough to justify the use of
one earth station.

Low Cost Interstate Voice: this service provides long distance telephone communication analogous to the WATS Dial Up facility.

Let us now describe briefly the charge structure for the CNS-A as representative of Integrated Communication Satellite Networks. SBS charges monthly for each Network Access Unit (the base earth station) and for the different Connection Arrangement Units (voice ports, data ports from 2.4 kbit/s to 1.5 Mbit/s) on each of the Network Access Units.

Network usage charges are split between charges for Full Time Transmission Units, which represent the capacity allocated to the customer network and Demand Transmission Units, which represent temporary overcapacity assigned to a network from the unallocated pool (see Fig. 2). FTU's are charged on a monthly basis while DTU's are charged on an hourly basis. To indicate an order of magnitude, a CNS-A network with three nodes with 25 voice ports per node and four 1.5 Mbit/s data ports, with permanent use of 25 FTU's (equivalent to 5.6 Mbit/s) costs in the area of 100 K$.

3. Conclusion

More integrated satellite communication networks will become operational in the next few years, not only in the United States but also in Europe where Telecom I, ECS and other projects are in active development phases.

As we are witnessing this tremendous explosion of the technology, the main challenge facing the European government, business and scientific community will be to find the proper balance between even accelerating technical progress, national interests and the need for controlled international communication.

DISCUSSION

Question: How do you solve the local network problem in the case of
shared earth stations?

Answer:	The same way as we do it with dedicated earth stations.
Shared stations are planned only in large cities or
business areas. The antenna will be on top of a high
building and we plan to cable it from there down up to a
few miles around. Right now there are no plans to add
micro wave links although that would probably be a good
solution.

Question: How many terminals are now connected to SBS, just to get
an idea about the size of the system?

Answer:	The satellite is not up yet, the satellite will be
launched October 27th, 1980. The plan is to have about 60
earth stations in 1981, and to increase that number up to
270 until 1984.

Question: I learnt from your presentation that you offer your service
for intra-company communication. Does this mean that if you
have customers who wish also to have inter-company communication that they need different terminals and different
network access units? How do you solve this problem?

Answer:	Right now SBS is planning to offer only the intra-company
communication and not to solve the inter-company communication problem. This is due to the FCC-policy.
From an interface point of view the SBS system is designed
to be transparent, so that as long you have V24 or RS 232
or same similar interfaces there will be no difficulties.
So if you will communicate to another network you will not
need another set of terminals. You will have a call from
one area to another and then have local connections within
the next area.

The New Communication Services:
Their Risks and Benefits for a Nation

W. A. McCrum
Ottawa, Ont./Canada

Summary

New forms of data communications services influence the social, economic
and strategic environment of a nation. The author discusses new forms of
data communications services from a Canadian perspective, and identifies
issues associated with the services and the new society they encourage.
The relationships of new services to national sovereignty, employment,
national trade balance, and regulation are discussed from a benefit-risk
perspective. The author concludes that a requirement exists for national
planning and coordination mechanisms to assure that appropriate systems
are in place to foster the orderly introduction, diffusion, and
utilization of new forms of data communications.

Zusammenfassung

Neue Formen der Datenkommunikation erleichtern das Heraufziehen der
Informationsgesellschaft und dabei gewinnen deren Einflüsse auf die
sozialen, wirtschaftlichen und außenpolitischen Strukturen einer Nation
an Bedeutung. Dieser Beitrag diskutiert die neuen Formen der Kommunikationsdienste aus kanadischer Sicht und stellt Beziehungen her zwischen
den Diensten und der neuen Gesellschaftsform, die dadurch gefördert
wird.

In einer Art Nutzen-Risiko-Analyse wird auf die Zusammenhänge zwischen
den neuen Diensten und der nationalen Unabhängigkeit, der nationalen
Handelsbilanz und der wirtschaftlichen Ausgewogenheit, der Beschäfti-
gungslage und der möglichen Steuermechanismen eingegangen. Der Beitrag
kommt zu dem Schluß, daß eine staatliche Planung und Koordination er-
forderlich ist, um sicherzustellen, daß angemessene Systeme zur Verfügung stehen, welche eine ordnungsgemäße Einführung, Verbreitung und
Nutzung der neuen Form der Datenkommunikation ermöglichen.

The growth of information-oriented societies driven by the continually diminishing cost of semiconductors presents problems of major concern to national governments of many countries, and particularly to developed Western countries, such as Canada. New communications services and systems are being announced with every new issue of communications trade publications.

Governments, with their own inertia against change resulting from their sheer size and complexity, are finding it extremely difficult to provide an adequate response in keeping with their important role of fostering the orderly development and operation of communications for their respective countries. In most cases the result has been a fragmented development of private communications networks and services with no overall framework to ensure compatibility between different implementations.

Such considerations make it timely and prudent to reflect on the risks and benefits of new communications services from a national perspective. The subject is one of significant interest to national governments and a great deal of time, money and effort is being expended to understand, predict, and ultimately control the development of new communications networks and services.

From the national point of view, it is expected that our conclusions, relating to North America, will in large measure be appropriate to the European scene since both communities share many common national developments and objectives.

<u>New data communications services in Canada</u>

<u>Service Providers</u>

The need for data communications services in Canada is met primarily by two nationwide competing common carriers, CNCP Telecommunications and the Trans-Canada Telephone System (TCTS). These organizations compete on a service-by-service basis for the domestic data communications market. In doing so, they have been exceptionally innovative in new data services development as is highlighted in the following section (Table 1).

<u>Table 1</u> - <u>Major data services in Canada</u>

<u>Service</u>	<u>Switching Technology</u>	<u>Speed Range</u>	<u>Carrier</u>
Direct distance dialled (DDD) network	Circuit-switched	Up to 1200 bit/s async Up to 2400 bit/s sync	TCTS
Telex	Circuit-switched	50 bit/s	CNCPT
TWX	Circuit-switched	Up to 110 bit/s	TCTS
INFODAT	Non-switched	Up to 56 Kbit/s	CNCPT
DATAROUTE	Non-switched	Up to 56 Kbit/s	TCTS
DATAPAC	Packet-switched	Up to 9.6 Kbit/s	TCTS
INFOSWITCH	Packet and circuit-switched	Up to 9.6 Kbit/s	CNCPT

<u>Service portfolios</u>

In the early 1970's Canadian carriers introduced leased-line synchronous digital data network services. These services were particularly cost-effective for the large volume data subscriber, besides providing significant improvements in transmission performance over the basic telephone network. This was followed in 1977 by the introduction of digital circuit and packet switching data services which provided further efficiencies in the allocation of network resources to the network subscriber with consequent cost savings.

The Canadian Datapac and Infoswitch networks are well known internationally, the former providing digital packet switching and the latter being a hybrid network providing both digital circuit and packet switching. These networks support a wide range of services, including support for synchronous packet mode intelligent terminals and support for asynchronous low-speed buffered and unbuffered terminals. This has permitted widespread application of the services to financial institutions, retail businesses, timeshare operations and a host of others.

Building on Datapac and Infoswitch technology, the network operators have now announced a wide range of new data communications services aimed at the automated office/information services market. These service portfolios include store-and-forward message switching, central word teleprocessing, computer conferencing, store-and-forward digital facsimile, text and graphics, forms processing and automation of manager/secretarial support functions.

In addition to this, two other developments are worthy of note. One is the joint project being conducted by the common carriers, cable television companies, and the Department of Communications to develop and field trial the Canadian Telidon system. Telidon is an advanced state-of-art videotex system, which has been widely announced and demonstrated. The impact of such a system on individual access to information would indeed be significant, if it were to achieve the widespread use that is anticipated.

The other major development of note in Canada is the evolving digital voice network. Plans are ready and implementation underway for introduction of a new family of digital multiplexing voice-switching machines developed by Northern Telecom, connecting to digital trunking facilities and eventually to digital local loops. This digital voice network will likely evolve to support a wide range of voice and data services. The resultant Integrated Services Digital Network (ISDN) is much discussed among the international telecommunications forecasting community.

Clearly, these briefly outlined developments bring with them many potential risks and benefits and certainly raise issues of concern to governments responsible for the orderly development and operation of communications within their national boundaries. In addition, the economic potential for exploitation of new communications technology by national industry is a matter of great interest to national governments. Consequently, many governments are now investing heavily in development of the new communications/information networks and services to increase the share of market for national industries and avoid the high cost of importing technology.

Problem areas

New forms of data communication facilitate the arrival of the information society, and give rise to a number of issues of strategic national importance. A multinational computer/data communications study revealed the following list of issues of concern: national sovereignty, vulnerability of the information society, privacy, employment, legality, and organizational operations.[1] In a study on the

impacts of the information society on Canada, concerns were identified relating to
the future of labour, the national and international location of information activities, implications for energy, the sociopolitical situation, and national vulnerability.[2] Finally, the 1979 Report of the Communications Research Advisory Board to
the Canadian Department of Communications listed the following issues related by
new communications services: employment. optimum industry structure, consumer interests, vulnerability, energy, and sovereignty. Many of these are surrounded by threats
and opportunities, by risks and benefits.

Risks and Benefits of new forms of data communication

Identifying the impact of precisely defined new forms of data communication
upon national economic, social and cultural environments is problematic. For example,
it is difficult to link a new videotex service to a change in the Gross National
Product (GNP) on a direct cause-effect basis.

The alternative is to view new forms of data communication as the vehicle
for increasing the information movement and automation of the information society;
this, in turn, can be related to macro-level national impacts. This is the alternative
we adopt here.

New forms of data communication, as with any technical change, can be
likened to a double-edged sword. The positive aspect relates to the creation of new
industries, services, and occupations increasing the efficiency and effectiveness of
the nation's organizations. The wide array of simple and convenient modes of conduct-
ing business and personal affairs made possible by new forms of data communication are
surely attractive benefits. The negative aspect relates to the demise of old industries, increased unemployment, and the replacement of old products and services.

National sovereignty

National sovereignty relates to the ability of a nation to control its
own destiny in an economic, technological and cultural sense. A nation is econo-
mically sovereign when it can make its own decisions on economic development and
resource deployment; it is technologically sovereign when it can assure that decisions related to business and industrial development are made primarily by its
nationals; a nation is culturally sovereign when its nationals can consume cultural
goods which are predominantly produced within its boundaries. A matter of particular relevance to national sovereignty is transborder data flow.

Transborder data flow is facilitated by new forms of data communications. That data may range from the innocuous, such as information on cloud formations, to corporate financial, personnel, administrative, research, planning, marketing, and evaluation data upon which foreign nationals may make decisions influencing another nation. The reaction to the foreign storage and processing of information has been labelled a 'massing haemorrhage of our independence' by Bernard Ostry, former Canadian Deputy Minister of Communications. Ostry goes on to state that this flow threatens Canada's "power to make decisions, to control our destiny. Indeed, with the power to make vital decisions located outside the country, we might discover that decisions affecting Canadians were being made subject to foreign laws. Private information about individual Canadian citizens could be in foreign hands.[3]"

Transborder data flow could weaken a nation's sovereignty and its ability to take the initiative directing the development of its economic, technological, and cultural development. The new forms of data communications technologies facilitating such transborder data flow enable the multinational corporation to coordinate the increasingly specialized activities of foreign branch plants and, if they so desire, to control the decision making of such branch operations.

Transborder data flow is of greater concern to those nations who include among their national industrial mix a large percentage of foreign-owned organizations. Such nations are particularly concerned because the outward information flow is often greater than the inward flow; because they can no longer control the impact of strikes and other types of service breakdowns; because they can no longer protect against damage, destruction, or misuse of the foreign-located information; because, once outside of a nation's borders, the use or misuse of information is ultra vires of the nation's legislative and legal system; and, finally, they are concerned with the number of jobs lost due to the foreign information processing made possible by new data communications services.

Privacy and security issues are raised by transborder data flow. Some time ago, Canadian newspapers recounted the story of New York City high-school students succeeding in tapping into the Trans Canada Telephone System's DATAPAC network, and actually destroying the files of a Canadian corporation. This incident illustrates the problems of the vulnerability of new data communications services and the privacy and security of the information they process and transmit.

The Swedish report Electronic Data Processing and the Vulnerability of Society[4] mentions the following vulnerability related issues: dependence on foreign countries, concentration of systems, dependence on a select trained operator staff, the sensitive nature of some information, terrorist activity, criminal action, threats, natural catastrophies, and accidents. Further, the Swedes found that, while protection against technical problems was 'high', protection against natural calamity was 'medium', and protection against premeditated deliberate attack was 'minimal'. The vulnerability of new forms of data communications takes on even more serious proportions between hostile nations.

A recurring issue is the impact of new forms of data communication on individual rights and privacy. Correspondence to the Canadian Privacy Commissioner, who is leading an investigation into the use of the Social Insurance Number, indicates that elements of the Canadian population fear that the Social Insurance Number is being used to link disparate files, thereby abusing personal privacy. What appears to be lacking, according to Commissioner Ingar Hansen, is the perception that files may be related in numerous other ways.[5] It is conceivable that innocuous information on one individual held in some 1500 Canadian federal databanks and numerous private databanks might be related to generate sensitive information which the individual would prefer not to be known. The Orwellian spectre of Big Brother emerges in this context.

The issue of misuse of financial information is of concern. Electronic funds transfer made possible by new forms of data communication facilitates the tracking of financial transactions both internationally and domestically. This could be a boon to unscrupulous credit companies, as such systems could provide means of recording a person's movements, the organizations to which he belongs and supports, his reading material, amusements, and patterns of consumption.[6]

Labour impact

Unemployment appears to be the major labour-related risk arising from the introduction of new communications services. As old jobs and roles in society disappear, new ones evolve to develop, market, implement, manage and evaluate new forms of data communications. There is no evidence to suggest that the number will be balanced. The transitional period between the disappearance of old roles and the appearance of new roles may prove volatile from a labour-management perspective.

The impact on employment will result from the increased concentration of capitalization in the services sector of the economy and the subsequent reduction of that sector's labour requirements. Unemployment would result from the increased productivity of the service sector, made possible by means of task substitution, process automation, intellect augmentation and process facilitation. Fewer individuals would be required to maintain the normal level of an organization's output. The impact on employment of increased capitalization of the service sector may prove substantial, as this sector has grown to comprise some 30 to 50% of many Western economies. The size of the reduction in labour requirements due to data communication, in particular, and computerization, in general, has been variously estimated as some 30% of the French labour force,[7] and 16% of the UK labour force.[8]

An undesireable byproduct of this anticipated reduction would be the closing of an escape valve for those found redundant in the primary and secondary sectors of the economy. These redundant personnel have traditionally migrated to the services sector in lieu of unemployment during economic downturns.

All is not dismal, however, for the employees of the service sector of the economy, for experience has demonstrated that the estimates of the negative impacts of new technology have been somewhat pessimistic. For example, the widely diffused automation scare of the 1950's and the 1960's never materialized, as automation diffused much more slowly than anticipated. Currently, the speed of diffusion of new data communication services is retarded by the shortage of skilled engineers, technicians, and a backlog of orders for silicon ships.[9]

Employment will of course, grow in the engineering, design, development, manufacturing, marketing, advertising, and servicing of new data communications technologies. It may also grow as a result of the competitive edge that can be gained by companies with the increased productivity made possible by new data communications services.

A major impact of the new forms of data communications and information technology will be the change in the mix of occupational roles. A change in skills will be required to allow the worker to interface with the new technologies; an understanding will be needed of the acquisition, processing, storage, retrieval and transmission of information by means of such technologies.

Thus, new data communications services contribute to unemployment, employment, and demand a new set of labour skills. This issue is succinctly summarized as follows by the Canadian Institute of Research on Public Policy.

"Whether the forthcoming revolution will create more jobs than it eliminates has been effectively argued both ways, and conclusions are more a matter of fear or faith than reason. However, there will be labour displacement, of that there can be no doubt. People will have to move entirely new kinds of employment and develop completely new skills and outlooks. This process will be lengthy and, unless new ideas about the distribution of wealth are considered, painful, anguishing and socially destructive.[10]"

The challenge is to ensure an orderly transition from the resource and skill requirements of today's workforce to those requirements being ushered in by new forms of communications.

Economics

New forms of data communication are facilitating the trend towards an information economy. An information economy exists when 50% of the GNP is produced by the information sector, wherein information includes all resources consumed in producing, processing, and distributing information goods and services. New data communication services facilitate the production, processing, and distribution of such information goods and services and, thereby, are a prerequisite for the information economy.

Data communication enables a nation to realize the benefits associated with the information economy by serving as a central nervous system for that economy, a system through which many of its transactions take place. Of particular economic concern is a nation's trade balance in the crucial areas of electronics and telecommunications.

Institutional and Regulatory Impacts

Institutional and Regulatory considerations can have significant impact on the risks and benefits of new forms of data communications. It is by no means obvious, for example, that a regulated monopolistic environment for provision of data communications will result in lower cost or better service to the end user than an open competitive environment. The arguments relating to cost advantages of broad based sharing of resources can be negated by removal of normal competitive incentives to keep prices down.

The opportunity exists for governments to establish a regulatory posture which will in fact relate to the new technology and new business infrastructures in a way which will have a positive national effect. One such suggestion in the 1979 Report of the Canadian Government Communications Research Advisory Board recommends "development of a clear definition of new service categories that are non-regulated and competitive, regulated but competitive, and regulated monopolistic." This would provide an environment of greater certainty for technological development.

The issue of carriage/content is one of importance for consideration by regulatory bodies and policy makers. Monopoly of carriage and content could limit the types and varieties of information available to the public and present the undesirable situation in which one national industry owns and operates an immensely powerful tool for manipulation of public attitudes.

Conclusions

New forms of data communications are transforming the workplace, the marketplace, and the home. Consequently, they are raising a host of problems relating to national sovereignty, employment, privacy, security, the economy and many other areas.

To maximize the benefits and minimize the risks for a nation, a number of basic needs have been identified. One is the need for a national policy for the coordinated development of new communications networks. Just as public highway systems require national direction and standards for their orderly development, so do the evolving complex and comprehensive telecommunications highways.

The need exists for a detailed understanding of services, and overall systems concepts, regarding their structures and economic application. In addition, national objectives with regard to industrial exploitation of new communications services are required to ensure that national industries derive major benefits as regards R&D, manufacturing and resultant international market development. The time for action is now: nations must exercise their economic, technological and cultural sovereignty.

REFERENCES

1. P.R. Robinson and L.A. Shackelton, National Policies and the
       Development of Automatic Data Processing, Department of
       Communications, Ottawa, 1979.

2. K. Valaskalsas, The Information Society: the Issues and the Choices,
       GAMMA, Montreal, 1979.

3. L. Dotto, 'Storing data outside Canada threatens freedom', Computer
       Data, May 1980, p. 42.

4. Ibid.

5. E. Ploman, 'Vulnerability in the information age', Intermedia, Vol 6,
       No 6, 1978, p 28.

6. M. Ryan and S. Baldwin, personal interview, 18 April 1980.

7. S.R. Hiltz and M. Turoff, The Network Nation: Human Communication
       via Computer, Addison-Wesley, Reading MA, 1978.

8. S. Nora and A. Minc. 'L'informatisation de la Societe', La Documenta-
       tion Francaise, Paris, 1978.

9. I. Barron and R.C. Curnow, The Future with Microelectronics:
       Forecasting the Effects of Information Technology,
       Frances Pinter, London, 1979.

10. C.A. Barron, 'Microelectronics: a survey', The Economist, 1 March
       1980, p. 3.

11. Russel, The Electronic Briefcase: the Office of the Future, Insitute
       for Research on Public Policy, Montreal, 1978.

12. 1979 Report of the Communications Research Advisory Board, Minister
       of Supply and Services, Ottawa, 1980.

13. J.M. McLean, The Impact of the Microelectronics Industry on the
       Canadian Economy, Institute for Research on Public Policy,
       Montreal, March 1979.

14. Op cit, Ref 12.

DISCUSSION after the lecture of A. McCrum
____

Question: How far is the development of higher protocols for the
          DATAPAC actually. What is the chance to get an open
          network?

Answer: The greatest activities are within the Canadian National
          Standards Body and the principal participants come from
          these two companies including the DATAPAC people. As far
          as the transport layer is concerned I believe that there
          is a national Canadian solution being developed only
          for the transport level protocol. Further activities
          have to do with the standardization of a file transfer
          protocol and the virtual terminal and the development of
          the Reference Model which will be so much smaller than
          the ISO-Model and more specific and being the model
          within which the Canadian Network will soon evolve.

Auswirkungen der neuen Kommunikationsdienste auf die herkömmliche verteilte Datenverarbeitung

A. Pott
Oberhausen

Zusammenfassung

Die Auswirkungen der neuen Kommunikationsdienste auf die Entwicklung der Verteilten Datenverarbeitung werden untersucht. Als Probleme werden näher behandelt der Trend zur Desintegration bezüglich Organisationsebene, Programmebene, Datenebene und Kommunikationsebene sowie die eingeschränkte Mobilität der Mitarbeiter. Bezüglich der Machbarkeit und Wirtschaftlichkeit verteilter Datenverarbeitungssysteme werden folgende Punkte näher untersucht: Standardisierung der Architektur verbundener Systeme, spezielle Eigenschaften verteilter Systeme, Kosten und Personalverfügbarkeit.

Zusammenfassend werden folgende Thesen begründet:
- Neue Kommunikationsdienste bieten eine Chance, die notwendige Standardisierung der Architektur von verbundenen Rechnersystemen zu fördern.
- Die neuen Kommunikationssysteme reduzieren den technischen und wirtschaftlichen Zwang, verteilte Datenverarbeitung einzuführen; sie sind für viele Anwendungsgebiete eine Alternative zur verteilten Datenverarbeitung.
- Neue Kommunikationssysteme sind weitgehend eine notwendige, aber keine hinreichende Voraussetzung für verteilte Datenverarbeitung. Für die graphische Informationsverarbeitung in verteilten Systemen sind die neuen Kommunikationsdienste praktisch eine conditio sine quo non.
- Die Integration von Faksimile- und Textübertragung sowie von Computer-Datenübertragung bedeutet einen Fortschritt für die Organisationsmöglichkeit der Unternehmen, und zwar insbesondere zur Eindämmung der möglichen Desintegration.

Summary

The impact of the new communications services on the development of
distributed systems is analysed. The paper deals with the trend
towards disintegration with respect to organization, programs, data
and communication relations and with limited personnel mobility.
There is a discussion on the feasibility and the financial aspects
of distributed systems. In this context the following points are
considered: standardization of the architecture of connected systems,
special features of distributed systems, cost aspects, and availability of personnel.

Arguments are given for the following four theses:

- New communications services give the chance to achieve the
  necessary standardization of the architecture of the connected
  systems.

- New communications systems reduce the technical and
  economical pressure for distributed processing. They form an
  adequate alternative for many applications.

- New communications systems are a necessary but not sufficient
  precondition for distributed systems. The new communications
  services turn out to be a conditio sine qua non for graphic
  information processing in distributed systems.

- The integration of facsimile-, text-, and data communication
  is of great importance for the development of new forms of
  organization in enterprises, especially with respect to
  potential disintegration.

Wie sich schon im bisherigen Verlauf der Tagung gezeigt hat,
sind die möglichen Auswirkungen der neuen Kommunikationsdienste
sehr vielfälig. Speziell die Auswirkungen auf die Entwicklung der
verteilten Datenverarbeitung sind ein vielschichtiges Thema, das
im Rahmen eines kurzen Vortrages leider nur unvollständig behandelt
werden kann. Mein Vortrag beschränkt sich daher auf die Behandlung
von einigen Einzelfragen.

Ihnen sind die neuen Kommunikationssysteme gestern und heute vorgestellt worden und ich kann mir ersparen, auf die Funktionen und Eigenschaften im einzelnen einzugehen. Weniger genau definiert und auf
dieser Tagung noch nicht besprochen sind verteilte Datenverarbeitungssysteme.

In meinem Vortrag gehe ich davon aus, daß es sich bei verteilten Datenverarbeitungssystemen um räumlich verteilte Datenverarbeitungssysteme
handelt, die durch Kommunikationssysteme miteinander verbunden sind.
Zwischen den räumlich verteilten Systemen werden Daten und Programme
ausgetauscht, d.h. die Verarbeitungsprozesse in den einzelnen System-
teilen kommunizieren untereinander und sind mehr oder weniger stark
gekoppelt. Ein Sonderfall der verteilten Datenverarbeitung ist die dezentralisierte Datenverarbeitung. In diesem Fall sind die Kommunikation
der Prozesse untereinander und die Kopplung gleich Null.
Für meine weiteren Überlegungen habe ich Annahmen über die Eigenschaften der neuen Kommunikationsdienste getroffen, die sich im wesentlichen
mit den Daten decken, die uns bisher bekannt geworden sind. Diese
Annahmen habe ich in einer kleinen Tabelle zusammengestellt:

. Effektive Übertragungsleistung etwa 256 Kbps
. Netzlaufzeit mit einer Wahrscheinlichkeit von
  99 % kleiner als 0.3 Sekunden
. Verfügbarkeit größer als 99.9 %
. Störungsdauer in 95 % der Fälle kleiner als 10 Minuten
. Anschlußgebühren für eine virtuelle Leitung kleiner
  als DM 200/Monat
. Mengenproportionale Kosten von weniger als 3×10^{-4} DM pro Paket
  von 64 Zeilen
. Unabhängigkeit der Übertragungskosten von der Entfernung.

Die aufgezeigten Leistungsdaten orientieren sich wie bereits gesagt
an den Planungszahlen der neuen Kommunikationsdienste, sind aber auch,
wie ich später noch aufzeigen werde, als Forderungen für den wirkungs-
vollen Einsatz der neuen Kommunikationsdienste in der Praxis anzusehen.

Ich werde bei einigen Überlegungen zur verteilten Datenverarbeitung
noch aufzeigen, welche Bedeutung diese Leistungszahlen für den Erfolg
bzw. für die Auswirkungen der neuen Kommunikationssysteme haben.

Der Ductus meines Vortrages ist etwas schwierig, weil ich aus der
Fülle der relevaten Problemfelder nur einige herausgreifen und behandeln kann. Um für Sie die Zusammenhänge meiner Überlegungen trotzdem
deutlich werden zu lassen, möchte ich zu Beginn einige Thesen formulieren, die ich im Verlauf des Vortrages durch die Analyse relevanter
Problemfelder näher begründen werden.

. Neue Kommunikationsdienste bieten eine Chance, die notwendige Standardisierung der Architektur von verbundenen Rechnersystemen zu
  fördern.

. Die neuen Kommunikationssysteme reduzieren den technischen und wirtschaftlichen Zwang, verteilte Datenverarbeitung einzuführen; sie sind
  für viele Anwendungsgebiete eine Alternative zur verteilten Datenverarbeitung.

. Neue Kommunikationssysteme sind weitgehend eine notwendige,
  aber keine hinreichende Voraussetzung für verteilte Datenverarbeitung. Für die grafische Informationsverarbeitung in verteilten Systemen sind die neuen Kommunikationsdienste praktisch
  eine conditio sine quo non.

. Die Integration von Faksimile- und Textübertragung sowie von
  Computer-Datenübertragung bedeutet einen Fortschritt für die
  Organisationsmöglichkeit der Unternehmen, und zwar insbesondere
  zur Eindämmung der möglichen Desintegration.

Bitte gestatten Sie mir zu Beginn einige Worte zur Euphorie über
die Möglichkeiten der verteilten Datenverarbeitung.

Sie wissen, daß die Datenverarbeitungsgemeinde sehr häufig dazu neigt,
Heilslehren zu formulieren. Der Fall, der uns allen sicherlich noch
vor Augen steht, ist die berühmte Lehre vom "Management-Informations-
System". Damals ging es darum, daß man sagte, man habe ein Rezept gefunden, um alle Manager glücklich zu machen, weil man einen Weg gefunden habe, ihre Informations-Probleme zu lösen. Bei der verteilten
Datenverarbeitung geht man nun einen kleinen Schritt weiter und man
sagt, man macht jetzt alle glücklich, denn man hat einen Weg gefunden,
wie die Schwächen der bisherigen Datenverarbeitung beseitigt werden
könnten.

Die häufigsten Argumente für eine verteilte Datenverarbeitung lauten
- wobei hier kein Anspruch auf Vollständigkeit erhoben wird - etwa
wie folgt:

. Die Hardware speziell für Microcomputer ist sehr viel billiger
  geworden; die Hardware-Kosten sind daher nicht länger eine Beschrän-
  kung für den Einsatz der Datenverarbeitung.

. Die Datenübermittlung ist zu teuer und zu langsam; Rechnerkapazität
  am Arbeitsplatz ist billiger und gibt bessere Antwortzeiten.

. Es gibt nicht genügend Programmierer, um den Bedarf an Programmen
  bei Benutzung zentraler Systeme zu decken.

. Die Microcomputer sind leicht zu handhaben und erweitern die
  Möglichkeiten ihrer Benutzer, insbesondere beseitigen sie alle
  Abhängigkeiten von zentralen Instanzen und motivieren die Mitarbeiter der Fachabteilung, den Einsatz der Datenverarbeitung
  zu erweitern.

. Verteilte Datenverarbeitungssysteme sind flexibler als zentrale
  Systeme.

Diese - wie ich noch zeigen werde - häufig sehr oberflächlichen Überlegungen dienen den Protagonisten der verteilten Datenverarbeitung als
Argument für eine möglichst weitgehende Verteilung von Rechnern an Endbenutzer. In einer der vielen Schriften wird mit Blick auf den Mangel
an geeigneten Programmierern sehr bildhaft formuliert: "Freuen wir
uns auf den Anblick, wenn die Endbenutzer die Ärmel hochkrempeln und
die Computer in Gang setzen". Man muß wohl von einem tiefen Glauben
durchdrungen sein, um an dieser Vision Freude zu haben; ich kann mir
kaum einen schlimmeren Anblick vorstellen.

Es bedarf keiner besonderen Vorstellungskraft, um zu erkennen, daß die
Verteilung der Datenverarbeitung beim heutigen Stand der Gestaltungs-
und Projekt-Management-Techniken und bei der völlig unzureichenden
Standardisierung einem modernen Turmbau zu Babel vergleichbar ist.
Dies gilt mit Sicherheit für die Fälle, wo in Organisationen isolierte
dezentrale Einzelrechner mit unterschiedlicher Systemarchitektur einge-
setzt werden. Dort wo mit unzureichenden personellen und technischen
Voraussetzungen der Ansatz zu verteilten Datenverarbeitungssystemen
gemacht wird, ist mit großer Wahrscheinlichkeit zu erwarten, daß diese
Systeme zu isolierten, dezentralen Systemen entarten.

Die Folgen einer Verteilung von DV-Kapazität ohne ein geschlossenes
Gesamtkonzept mit geeigneten Durchsetzungs- und Kontrollmöglichkeiten
können für die betroffenen Organisationen sehr weitreichend sein.
Von den auftretenden Problemen möchte ich einige, die einen Bezug

zum Thema Neue Kommunikationssysteme und Verteilte Datenverarbeitung
haben, besonders herausstellen:

- Trend zur Desintegration
- Eingeschränkte Mobilität der Mitarbeiter

Um die Auswirkungen der neuen Kommunikationssysteme auf die verteilte
Datenverarbeitung zu behandeln, sind neben diesen mehr unternehmenspolitischen Themen andere Fragen zu behandeln, die die Machbarkeit verteilter Datenverarbeitung betreffen und das Thema der Wirtschaftlichkeit einer solchen Strategie näher beleuchten. Im einzelnen möchte ich
daher auf folgende Punkte eingehen:

- Standardisierung der Architektur verbundener Rechnersysteme

- Spezielle Eigenschaften verteilter Datenverarbeitungssysteme

- Kosten und Personalverfügbarkeit.

Da die Behandlung der zuletzt genannten Problemfelder das Verständnis
der unternehmenspolitischen Probleme erleichtert, werde ich diese
Problemfelder zuerst behandeln.

Sie werden unschwer bemerkt haben, daß ich bezüglich der Einführung
verteilter Datenverarbeitungssysteme sehr zurückhaltend bin. Um Mißverständnisse zu vermeiden, möchte ich bereits zu Beginn feststellen,
daß ich keinen Glaubenskrieg gegen die verteilte Datenverarbeitung
führe. Es geht nach meiner Auffassung nicht darum, _ob_ verteilte Datenverarbeitung eingeführt werden wird sondern _wie_ und _wann_ dies mit ab-
schätzbaren Entwicklungsrisiken, vertretbarem Aufwand und einer hin-
reichenden Kenntnis der Wirkungen im unternehmenspolitischen und
sozialen Bereich geschehen kann.

Die zu frühe Proklamation von theoretisch denkbaren DV-Konzepten
als Problemlösungen für Tagesprobleme, die falsche Einschätzung
von Schwierigkeiten, Aufwand, Akzeptanz und Wirkung solcher Konzepte
haben dem Ruf der Datenverarbeitung sehr geschadet. Die wohl bedachte
Bewältigung der nächsten Entwicklungsstufe, zu der die verteilte Datenverarbeitung und die neuen Kommunikationssysteme gehören, kann dazu
beitragen, große Schäden zu vermeiden und verlorenes Terrain zurück-
zugewinnen.

Standardisierung der Architektur verbundener Rechnersysteme

Gegenstand der Betrachtung sind Systeme, die aus einer Vielzahl
von untereinander verbundenen Einzelsystemen bestehen. Die neuen
Kommunikationsdienste sind eines der Teilsysteme der verteilten
Datenverarbeitung.

Die bestehenden Architekturen verteilter Datenverarbeitungssysteme
sind gewachsen und haben sich bei den einzelnen Herstellern aus der
jeweils verfügbaren Hardware und Software entwickelt. Das Ergebnis
dieser Evolution ist eine Vielfalt von inkompatiblen Architekturen.

Der normale Anwender ist quasi in diesen Architekturen gefangen, da
der Aufwand für Systemlösungen, die zwei oder mehr Architekturen
implizieren, nur mit außerordentlichem Aufwand realisiert werden
können; die Verschiedenheit der Architektur schafft eine starke Bin-
dung der Anwender an den Hersteller. Ohne boshaft zu sein, kann man
sicherlich sagen, daß dies einer der Gründe für die geringen Fortschritte bei der Standardisierung ist.

Die vorgetragenen Ansätze für neue Kommunikationsdienste sehen in
unterschiedlichem Maße vor, diese mit der Eigenschaft eines
"transparent pipeline-service" auszustatten. Ich halte diese Entwicklung aus Anwendersicht für fragwürdig. Abgesehen davon, daß bei
den notwendigen Transformationen zwischen den Architekturen, d.h. den
Protokollen und sonstigen Schnittstellen der Einzelsysteme, Abbildungs-
verluste auftreten, die den Lösungsraum der beteiligten Architektur
einengen und, auch abgesehen davon, daß für die Realisierung dieser
Funktionen ein erheblicher Aufwand bei der Entwicklung und im Betrieb
entsteht, ist die Entwicklung des transparent pipeline-service nach
meiner Meinung ein Schritt in die falsche Richtung.
Die Möglichkeit, in der Kommunikationssphäre die bestehenden Architekturen weiter zu verwenden, begünstigt die gewachsene unnötige Architekturvielfalt, ohne daß das Problem der Inkompatibilität zwischen den
einzelnen Architekturen gelöst wird. Die Entwicklung geeigneter
Standards ist unvergleichlich wirtschaftlicher und läßt nach meiner
Auffassung einen ausreichenden Spielraum für den Wettbewerb der Hersteller, der sich auch unterhalb einer einheitlichen Benutzer-Schnitt-
stelle abspielen könnte.

Viele Probleme in der Datenverarbeitung sind eine Folge des Überflusses
an Lösungsmöglichkeiten. Mit Sicherheit sind im Anwendungsbereich nur

sehr selten Schwierigkeiten anzutreffen, weil das verfügbare technische Instrumentarium die wirklich notwendigen Lösungen nicht zuläßt.

Die aus der Architekturvielfalt resultierenden Probleme der Anwender, auf die ich bei der Behandlung der weiteren Punkte noch eingehen werde, sind so groß, daß es mir überlegenswert erscheint, in den nächsten Jahren bewußt Abstinenz bei der weiteren Einführung der Datenverarbeitung zu üben. Soweit ich es übersehen kann, ist dies der wirkungsvollste Beitrag, den die bisher in allen strategischen Fragen sehr zurückhaltenden Anwender leisten könnten, um die Standardisierung voranzutreiben. Die neuen Kommunikationssysteme können bei dieser Entwicklung in Richtung auf eine Standardisierung eine außerordentliche Rolle spielen, da dieses zentrale Teilsystem der verteilten Datenverarbeitung für die weitere Entwicklung der Standardisierung ein sehr geeigneter Ansatzpunkt ist.

Wie ich bereits erwähnt habe und noch zeigen werde, sind viele Leistungsmerkmale der neuen Kommunikationssysteme notwendig für die erfolgreiche Einführung der verteilten Datenverarbeitung. Ohne die Standardisierung der Architektur verbundener Rechnersysteme, d.h. ohne ein Konzept, das die Risiken, den Aufwand und den Entscheidungsspielraum der Anwender wirklich berücksichtigt, ist die Durchsetzung der verteilten Datenverarbeitung und der neuen Kommunikationsdienste - besonders wenn diese nicht von einem etablierten DV-Hersteller angeboten werden - sehr fraglich.
Nach meiner Meinung ist es eine bedenkenswerte Alternative, Geld nicht in "transparent pipelines" (wo führen diese hin?) sondern in die "realen Kanäle", die die ISO-Bemühungen vorantreiben, zu pumpen.

Kosten und Personalverfügbarkeit

In der Praxis der Auseinandersetzung mit der verteilten Datenverarbeitung bestimmen, ohne daß dies immer gerechtfertigt wäre, fast nur die sogenannten Kostenvergleiche eine Rolle. Sogenannte, weil meist nur unmittelbare Kosten und Ausgaben - und die nicht einmal vollständig - in Ansatz gebracht werden, während alle Risiken und mittelbaren Auswirkungen der Planung außer Ansatz bleiben.

Entscheidende Argumente der Protagonisten verteilter Datenverarbeitung sind die Kosten der Datenübertragung bei Benutzung zentraler Rechenzentren. Die Leistungsangebote der Post reichen, bis auf einige Ausnahmen im Bereich der grafischen Datenverarbeitung, aus.

Die grafische Datenverarbeitung wird allerdings in den nächsten
Jahren eine wesentlich größere Verbreitung finden, speziell im
Bereich des computer aided design (CAD). Hier sind die Anforderungen
bezüglich der Übertragungsleistungen etwa um den Faktur 10 höher als
bei normaler Verarbeitung im Dialogmodus. Die bezahlbaren Übertragungsleistungen, d.h. 1,2 - 9,6 KBd, sind in vielen Fällen nicht ausreichend,
insbesondere treten aufgrund von Warteschlangen starke Streuungen der
Übertragungsleistungen auf. In diesem Bereich ist eine starke Steuerung der Übertragungsleistung aber außerordentlich störend, da sie
sich mit der Steuerung der Wartezeiten und der Verarbeitungszeiten im
Rechner überlagert.

Bild 1 zeigt eine typische Streuung der Übertragungsleistung bei
einer CAD-Anwendung in einem 250 km vom zentralen Rechenzentrum ent-
fernt gelegenen Werk.

Um zufriedenstellende Übertragungsleistungen zu erreichen, d.h. die
Argumente bezüglich technischer Unzulänglichkeiten zu entkräften,
müßten schnellere Leitungen angemietet werden. Dies würde bei heutigen
Preisen zu Leitungskosten führen, die erheblich höher wären als die
Mietkosten für das zentrale Rechnersystem.

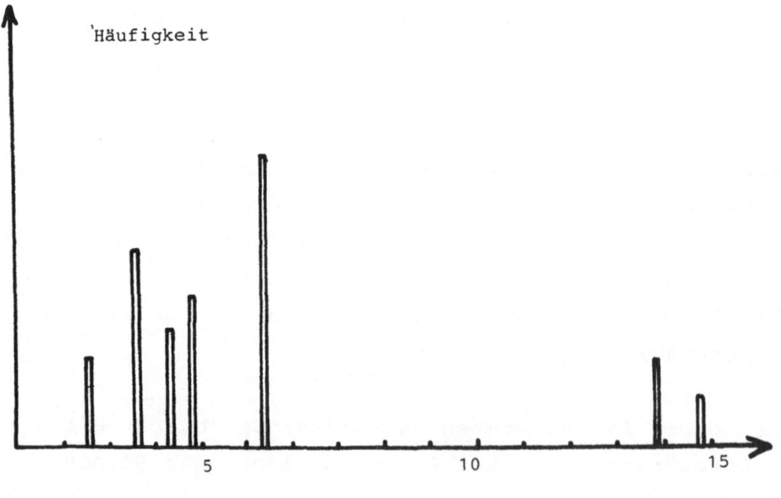

Bild 1

In allen Fällen, in denen die geforderte Übertragungsspitzenleistung
sehr hoch ist und/oder die mittlere Auslastung der Leitung gering ist,
sind die Kosten für die Übertragungskomponenten so hoch, daß eine
Verteilung der Datenverarbeitung aus Kostengesichtspunkten gefordert
wird. Es ist eine triviale Erkenntnis, daß neue Kommunikationssysteme
mit den zu Beginn genannten Preis- und Leistungskenndaten diese Lage
wesentlich verbessern werden und so die technisch und konzeptional
unnötige Verteilung von Datenverarbeitung mit allen ihren Problemen
vermieden werden kann.

In der Sphäre der Kostenüberlegungen gehört auch die Betrachtung
der Personalsituation. Es gibt hier ein quasi technisches Problem,
das durch den Mangel an ausreichend qualifiziertem Personal begründet ist: die fehlende Qualifikation schränkt die realen Möglichkeiten
der Systementwicklung erheblich ein und verursacht häufig sehr hohe
Kosten.
Der Mangel an qualifiziertem Personal führt außerdem zu außerordent-
lichen Steigerungen der Personalkosten.

Es ist nach meiner Auffassung ein folgenschwerer Irrtum anzunehmen,
die Verteilung von Datenverarbeitung wäre ein Weg, die fehlende
Qualifikation im personellen Bereich der Datenverarbeitung zu kompen-
sieren. Indem Personal aus Fachabteilungen in großer Zahl nach einer
Schulung zur Programmierung von Computern eingesetzt wird, ist vielleicht ein Mengenproblem gelöst, mit Sicherheit aber nicht das Qualifi-
kationsproblem. Die Lernfähigkeit wird im allgemeinen stark überschätzt.
Ich glaube, daß man davon ausgehen muß, daß die Zahl der qualifizierten
DV-Anwender in den Fachabteilungen auch in Zukunft sehr begrenzt sein
wird. Zum anderen wird der Entwurf, die Entwicklung, die Wartung und
die spätere Anpassung an betriebliche Veränderungen bei verteilten
Datenverarbeitungssystemen erheblich komplexer und damit schwieriger
als bei den heutigen "zentralen", d.h. nicht verbundenen Rechnersystemen. Dieser Zuwachs an Schwierigkeiten, der ein verdecktes Kostenpoten-
tial ist, resultiert u.a. aus dem schon besprochenen Fehlen von Standards für die Architektur verteilter Systeme, der daraus resultierenden
Änderungsrate der Architektur und aus dem Zwang, in verteilten Datenverarbeitungssystemen lastabhängig die Datei- und Programmverteilung
zu ändern, um die benötigten Antwortzeiten zu erhalten.

Diese Überlegungen beziehen sich allerdings nicht auf dezentralisierte
und isolierte Einzelsysteme, deren Problematik auf anderen Gebieten
liegt.

Die Konsequenz aus der geschilderten Situation ist die Notwendigkeit,
mit dem verfügbaren Personal sehr rationell zu arbeiten. Ein Weg dies
zu tun, ist die Konzentration der Spezialisten in Zentralen, um von
dort aus alle zentralisierbaren Funktionen in Netzen von Rechnersystemen zu betreuen, wozu neben der Wartung von Hardware und Betriebs-
software auch das Bedienen verteilt installierter Rechner gehören.

Es ist ein Fehler, heute im Zuge der voreiligen Dezentralisierung
von Datenverarbeitungssystemen Arbeitsplätze für Bedienungs- und
Betreuungsaufgaben zu schaffen, deren Überflüssigkeit bereits nach
sehr kurzer Zeit zu ihrer Abschaffung führen wird. Der Drang man-
cher Datenverarbeiter zur "Selbständigkeit" zeigt hier bedenkliche
Auswirkungen.

Je eher neue Kommunikationsdienste in dieser Lage einen Beitrag leisten,
durch ihre Übertragungsgeschwindigkeit und ihre Sicherheit die Voraussetzungen für ein zentral betreutes verteiltes Datenverarbeitungsnetz
zu schaffen, um so schneller können Fehlentwicklungen durch eine
richtige Strategie ersetzt werden.

Um Mißverständnisse zu vermeiden, möchte ich an dieser Stelle betonen,
daß die hier aufgezeigten zentralen Dienstleistungen die organisatorischen Gestaltungsmöglichkeiten und die Freiheit der Nutzung von
DV-Leistungen nicht einschränken. Die Verteilung von Datenverarbeitungssystemen muß aber nicht nur aus personellen und Kostengründen, sondern
aus systemtechnischen Gründen als ein geschlossenes System geplant
und von einer zentralen Instanz betrieben werden, um arbeitsfähig
zu sein. Ohne diesen Punkt wirklich behandeln zu können, möchte ich
auf einige der speziellen Eigenschaften von verteilten Datenverarbeitungssystemen kurz eingehen.

Spezielle Eigenschaften verteilter Datenverarbeitungssysteme

Definitionsgemäß handelt es sich bei der verteilten Datenverarbeitung
um kommunizierende Prozesse, die auf räumlich verteilten Systemen ab-
laufen. Für das Design verteilter Systeme ist die Kopplung der Prozesse untereinander von grundlegender Bedeutung. Ohne auf die Definition des Begriffes Kopplung an dieser Stelle näher einzugehen - was
ohnehin schwierig wäre, da die theoretische Diskussion dieses Problems
noch nicht abgeschlossen ist - sei hier der Einfachheit halber ange-

nommen, daß die Kopplung bestimmt ist durch die Größe der "Schlupfzeit"
zwischen abhängigen Prozessen. Die "Schlupfzeit" ist die Zeit, die
einem Prozeß nach Eintreffen aller relevanten Informationen zur Verfügung steht, bevor er beginnen muß, um zeitgerecht zu einem Ergebnis
zu kommen. Die Kopplung ist umso geringer, je größer die frei verfügbare Zeit (Schlupfzeit) ist. Für die Übertragung der notwendigen Informationen zwischen zwei Prozessen wird Zeit benötigt, die von der
Leistung des Kommunikationssystems bestimmt wird. Je schneller das
Kommunikationssystem arbeitet, umso geringer wird bei sonst gleichen
Bedingungen die Kopplung.
Die Gestaltungsmöglichkeiten von verteilten Datenverarbeitungssystemen
sind erheblich größer, wenn es gelingt, die Kopplung der Prozesse
zu reduzieren. Besondere Bedeutung hat die geringe Kopplung für die
Kohärenz des Systems, die Konsistenz der verteilten Daten, die Ver-
meidung von Überlastungssituationen oder auch die Konvergenz des
Systemzustandes nach Störungen.

Die hier angesprochenen Eigenschaften verteilter Datenverarbeitungssysteme sind zusammen mit anderen Zielgrößen die Kostenminimierung
und Antwortzeitverhalten auf komplexe Weise verknüpft. Die Erörterung
dieser Zusammenhänge und der jeweiligen Auswirkung neuer Kommunikationsdienste auf einzelne Eigenschaften und Zielgrößen der Systeme würde
den Rahmen dieses Referates sprengen. Eine pauschale Betrachtung
läßt jedoch bereits die Zusammenhänge erkennen.

Da ein Teil der Schlupfzeit bei Auftreten von Fehlern oder durch
eine fallweise Verlängerung der Übertragungszeit (Streuung) ver-
braucht wird, ist unmittelbar erkennbar, daß die Übertragungsgeschwindigkeit und die Sicherheit des Übertragungssystems ebenso
wie die geringe Streuung der Übertragungszeiten von sehr großer
Bedeutung für die Realisierung verteilter Datenverarbeitungssysteme
sind. In vielen Fällen dürften die neuen Kommunikationsdienste mit
ihren hohen Leistungsdaten bezüglich Übertragungsgeschwindigkeiten
und Sicherheit sogar eine unverzichtbare Voraussetzung für die
Realisierung verteilter Datenverarbeitungssysteme sein, speziell
wenn es sich um Systeme mit hohen Anforderungen an die Zeitgenauig-
keit der Prozesse handelt.

Nach der kurzen Behandlung der Fragen, die mehr zur eigentlichen
Datenverarbeitungssphäre gehören, möchte ich noch auf einige Punkte
eingehen, die die unternehmenspolitische Relevanz des Themenkomplexes
deutlich werden lassen. Die Interdependenz von speziellen Systemeigen-
schaften und Entwicklungen in der Anwendersphäre wird dabei deutlich.

Trend zur Desintegration

Wie ich bereits angedeutet habe, ist ein Sonderfall der verteilten
Datenverarbeitung die sogenannte dezentrale Datenverarbeitung, die
ich jedoch in Zukunft als "isolierte Datenverarbeitung" bezeichnen
möchte. Die Isolierung eines Datenverarbeitungssystems läßt sich
auf verschiedenen Ebenen betrachten. Für unsere Überlegungen kann
man von vier Ebenen oder Konsistenzbereichen ausgehen:

- Organisationsebene
- Programmebene
- Datenebene
- Kommunikationsebene.

Die Isolierung eines Systems bzw. Teilsystems ist umso ausgeprägter,
je größer die Inkonsistenz in dem jeweiligen Bereich ist.

Die Funktionsfähigkeit eines arbeitsteilig organisierten Unternehmens
hängt davon ab, daß in allen Ebenen eine möglichst große Konsistenz
gegeben ist. Um diese Aussage etwas zu erläutern, möchte ich nur ein
Beispiel für die Datenkonsistenz erwähnen: Es ist ohne weiteres einzu-
sehen, daß die in einem Unternehmen vorhandenen Auftragsdaten und
Adressdaten, gleichviel in welchem Unternehmensteil sie benutzt werden,
konsistent sein müssen, um insgesamt zu richtigen Arbeitsergebnissen
zu kommen.

Für die Behandlung des von mir angesprochenen Phänomens muß über den
Einfluß der Komplexität gesprochen werden. Wenn man, was der Einfachheit halber hier gestattet sei, davon ausgeht, daß die Komplexität
in erster Näherung proportional zur Anzahl der auf allen Systemebenen
vorhandenen Elemente und Verknüpfungen ist, dann läßt sich leicht ab-
leiten, daß ein Aufwandsminimum für einzelne Arbeitsergebnisse erzielt
werden kann, wenn die Zahl der Verknüpfungen gesenkt wird, d.h. wenn
die einzelnen Aufgaben stärker isoliert werden. Für unsere Überlegungen
kann man weiterhin die Annahme treffen, daß in einem Unternehmen die
Verknüpfungen von Daten, Algorithmen und sonstigen Regelungen vorgegeben und ein konstitutives Merkmal des Unternehmensorganismus sind.

Die interlektuellen Schwierigkeiten und der Aufwand bei der Behandlung
komplexer Zusammenhänge führen dazu, daß in Organisationsstrukturen
ständig Bestrebungen wirksam sind, den Grad der Komplexität zu verrin-
gern. Wenn man, wie bereits gesagt, davon ausgehen muß, daß die Komplexität bei unverändertem Unternehmensaufbau und Unternehmenszweck

weitgehend invariant ist, die meisten Verknüpfungen für die Gesamt-
funktion also notwendig sind, bedeutet die Verminderung der Komplexität
eine zunehmende Inkonsistenz und damit eine Desintegration der Unter-
nehmensfunktionen.

Einige Eigenschaften von Datenverarbeitungssystemen begünstigen
Entwicklungen in Richtung auf eine zunehmende Desintegration.
Hierzu muß kurz auf die Zusammenhänge zwischen Aufwand und Kom-
plexität bei Datenverarbeitungssystemen eingegangen werden.

Es wird sicherlich nicht bestritten, daß die Erstellung von Arbeitsergebnissen in einer komplexen Systemumgebung aufwendiger ist als
dies bei einfachen Systemen der Fall ist. In Bild 2 ist in einer
einfachen Grafik dargestellt, wie sich bei Rechnern verschiedener
Größenklassen der Aufwand in Abhängigkeit von der Komplexität der
Systemumgebung verhält. Das Bild zeigt sehr deutlich, daß Mikrocom-
puter in einer einfachen Umgebung bei einfachen Problemen sehr viel
geringeren Aufwand für die Leistungserstellung benötigen, als dies
bei größeren Rechnern der Fall ist.

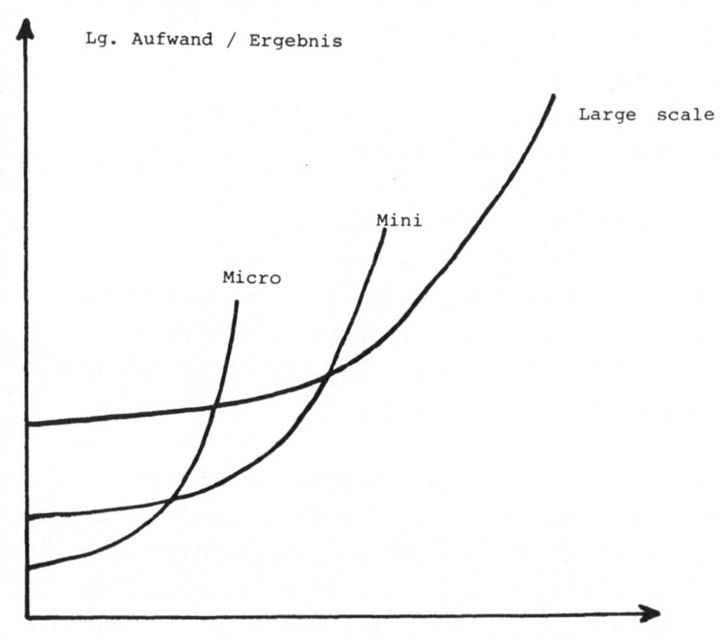

Bild 2

Die Einführung von Datenverarbeitungssystemen geschieht fast aus-
nahmslos auf die Weise, daß zu Beginn ein Einzelproblem mit Hilfe
eines Datenverarbeitungssystems gelöst werden soll. Die Anfangs-
situation ist daher in aller Regel mit dem Fall eines nicht komplexen Systems identisch.

Diese Überlegungen zeigen, warum eine Verteilung von Datenverarbeitung in der Anfangsphase nur geringe Probleme aufwirft, und zwar um
so mehr, je mehr einzelne Systeme für die Lösung isolierter Aufgaben
eingesetzt werden; im Grenzfall ein System je Aufgaben.

Der Markt wird zur Zeit von spezialisierten Systemen, die in dem
hier besprochenen Sinn den Charakter von isolierten Systemen haben,
überschwemmt. Ich darf Sie an Schreibautomaten, Karteicomputer,
Terminplanungs-Rechner und alle die vielen anderen Geräte erinnern.
Es sieht so aus, als sei der "Siegeszug" verteilter Datenverarbeitung
nicht mehr aufzuhalten.

Betrachtet man die Aufwandsentwicklung in Bild 2, so zeigt sich
unmittelbar, daß bei einer Integration der Einzelaufgaben unter
zunehmender Berücksichtigung der notwendigen Verknüpfungen der Aufwand außerordentlich stark ansteigt. Dieser starke Anstieg ist der
Grund dafür, daß Konzeptionen von verteilten Datenverarbeitungssystemen
nach ersten "Erfolgen" aus Mangel an Personal und Sachmitteln zu isolierten Systemen entarten bzw. in diesem Zustand bleiben. Insbesondere
die oft für verteilte Datenverarbeitung reklamierte Flexibilität,
die zu ihrer Realisierung hervorragende Systementwürfe benötigt,
ist in der Regel das erste Opfer des Unvermögens.

Die unbedingt notwendige Konsistenz von Daten und Programmen wird
häufig über "manuelle Brücken" hergestellt, wodurch ein schwer er-
faßbarer Aufwand entsteht, der nach meiner Meinung in den meisten
Fällen die Rationalisierungsergebnisse der einzelnen Teilsysteme
übersteigt; ein Effekt, der bereits bei vielen konventionellen Datenverarbeitungssystemen beobachtet werden kann.

Die bisherigen Überlegungen zeigen, daß die beobachtete Entwicklung
zum "Schreibtischcomputer", zu Systemen also, die für einzelne Arbeitsplätze meist isolierte Aufgaben lösen, theoretisch leicht erklärbar
ist und daß eine vordergründige Wirtschaftlichkeitsrechnung zu dem
Ergebnis führen muß, daß die Verteilung von Datenverarbeitung eine
richtige Strategie ist.

Eines der Probleme liegt darin, daß die Komplexität nicht ohne
weiteres erkennbar ist und in aller Regel keine Notation vorliegt,
aus der Erkenntnisse über den Grad der Verknüpfung gewonnen werden
können. Entsprechend ist die Entwicklung einer Desintegration in kleinen Schritten nur sehr schwer festzustellen.
Die frühesten Symptome einer Desintegration zeigen sich oft durch
eine Abnahme der Flexibilität, die dadurch bedingt ist, daß isolierte
Systeme in der Regel auf wenigen festgeschriebenen Umweltbeziehungen
aufbauen, die sich praktisch nicht verändern lassen. Dieses Phänomen
ist außerordentlich interessant und von großer Bedeutung für die Entwicklung der nächsten Jahre. Die Behandlung sprengt jedoch den Rahmen
dieses Referates.

Wie ich zu Beginn schon gesagt habe, können die neuen Kommunikationsdienste durch einige ihrer Funktionen nicht nur neue Möglichkeiten
der Organisation schaffen, d.h. im positiven Sinne den Aufbau integrierter Informationssysteme fördern sondern auch der geschilderten
Desintegration entgegenwirken. Besonders gilt dies selbstverständlich
für die Kommunikationsebene. Jedoch sollte die Integration von Computer-
Daten, Textdaten und Faksimile-Übertragung einen starken Impuls auch
für die Integration in den anderen Ebenen ergeben. Je mehr es gelingt,
die Vielzahl der Hersteller von Spezialsystemen durch die Anschluß-
möglichkeit an Kommunikationsdienste zu einer Akzeptanz von allgemein
verbindlichen Standards zu bewegen, umso größer dürfte der angestrebte
Effekt sein.

Eingeschränkte Mobilität der Mitarbeiter

Für die Wirtschaft ist die Mobilität der Mitarbeiter - womit hier die
Breite des möglichen Einsatzspektrums gemeint ist - in zunehmendem
Maße ein wichtiger Faktor, sich des Unternehmensumfeldes anpassen zu
können. Die Benutzeroberfläche von Datenverarbeitungssystemen für End-
benutzer bedeutet für das Personal der Fachabteilungen ein zusätzliches
Bündel von Kenntnissen und Fähigkeiten, die zudem in der Regel schwie-
riger zu erlernen sind als der Umgang mit klassischen Informations-
und Kommunikationsmitteln. Kenntnisse und Fähigkeiten, die für einen
Arbeitsplatz benötigt werden, sind ein relevanter Sachverhalt des
Betriebsverfassungsgesetzes (BVG). Änderungen der diesbezüglichen
Anforderungen erfüllen den Tatbestand einer Versetzung, die nach dem
BVG mitbestimmungspflichtig ist.

In dem Maße, wie Kenntnisse und Fähigkeiten für die Nutzung der Datenverarbeitung in einem Unternehmen arbeitsplatzspezifisch sind, sinkt
die Mobilität der Mitarbeiter, da zur Veränderung der fachlichen
Anforderungen bei Umsetzungen DV-spezifische Änderungen hinzukommen.
In Fällen, wo bei gleichen Funktionen bzw. Aufgaben mit entsprechend
gleichen technischen Anforderungen unterschiedliche Datenverarbeitungssysteme eingesetzt sind, wird die negative Auswirkung der Datenverarbeitung besonders deutlich. Wer die Fakturierung auf einem, sagen wir
Siemens-System beherrscht, kann nicht ohne weiteres die gleichen Arbeiten mit einem DEC-System durchführen! Interessant in diesem Zusammenhang ist auch die Überlegung, daß bei unveränderten Aufgaben
am gleichen Arbeitsplatz allein durch die Änderung des verwendeten
Datenverarbeitungssystems nach dem BVG ein mitbestimmungspflichtiger
Vorgang - nämlich eine Versetzung - gegeben ist.
Für die Einführung von Datenverarbeitungssystemen an Arbeitsplätzen
in Fachabteilungen ist es daher eine unverzichtbare Forderung, daß der
Katalog der für die Benutzung der Datenverarbeitung notwendigen Kenntnisse und Fähigkeiten einheitlich, möglichst klein und zeitlich langfristig konstant ist.
Die Forderung, daß die Kenntnisse und Fähigkeiten Hersteller-unabhängig
sein müssen, möchte ich noch einmal ausdrücklich erwähnen, obschon die
bereits genannten Forderungen dies implizieren.

Nach diesen Vorbemerkungen wird deutlich, welche Bedeutung eine Verteilung von Datenverarbeitungssystemen zukommt, wenn diese dazu führt,
daß isolierte inkompatible Datenverarbeitungssysteme eingeführt und
so Arbeitsplätze einer anderen Qualität geschaffen werden. Die Apsekte,
die bereits bei der Desintegration behandelt wurden, bekommen hier eine
zusätzliche Bedeutung. Insbesondere fällt auf, daß die Standardisierung
im Zusammenhang mit der verteilten Datenverarbeitung eine wesentlich
größere Bedeutung hat, als dies bisher ohnehin schon der Fall war.
Jeder Beitrag, den die neuen Kommunikationsdienste leisten,
um die Standardisierung weiter zu entwickeln, ist ein wichtiger
Schritt, um akzeptable Voraussetzungen für die verteilte Datenverarbeitung zu schaffen.

Es ist nicht leicht, in einem kurzen Vortrag die Auswirkungen
der neuen Kommunikationsdienste auf die herkömmliche verteilte
Datenverarbeitung erschöpfend zu behandeln. Der Nutzen der neuen
Kommunikationsdienste ist nach allem, was bisher besprochen wurde,
allerdings offensichtlich, wobei aus Anwendersicht einige aufmun-
ternd kritische Bemerkungen gemacht werden mußten.

Auswirkungen der neuen Kommunikationstechniken auf Struktur und Ablauf von Tätigkeiten in einem Unternehmen

B. Czaputa
München

Zusammenfassung

In einem Großunternehmen wächst stetig der Bedarf zur weltweiten Kommunikation. Die Verbesserung des Kommunikationssystems in allen seinen Ausprägungen erfordert organisatorische, personelle wie auch technische Investitionen.

Mit dem Projekt SIEKOM, mit dem das innerbetriebliche, überbereichliche "Kommunikationssystem" verbessert werden soll, werden Vorgehensweise und personelle Maßnahmen aufgezeigt. Technische Aspekte werden in Anbetracht der vielfältigen Beiträge der anderen Referenten bewußt nicht behandelt.

Summary

In a large enterprise there is a permanently growing need for worldwide communications. The necessary improvement of the employed communication system with its various aspects requires huge capital investments with respect to organization, personnel and technology.

The project SIEKOM is to improve the intra-company worldwide communication. The paper explains the intention, the measures to be taken and activities in the personnel area. Technical aspects of the communication system are not dealt with since they are addressed by the other contributions to the conference.

1. Einleitung

Seit 10 Jahren trage ich Verantwortung für Organisation und Automatisierung in einem Teilbereich des Hauses Siemens. Darüberhinaus bin ich seit einiger Zeit Projektleiter für den Ausbau der innerbetrieblichen Einrichtungen zur weltweiten Kommunikation.

Bei meinen Ausführungen bitte ich daher zweierlei zu berücksichtigen: Erstens behandle ich das Thema aus der Sicht eines Großunternehmens, das weltweit tätig ist. Damit sind verschiedene Statements zu relativieren, wenn sie mit Verhältnissen anderer Branchen oder kleinerer Unternehmen verglichen werden.

Zweitens spreche ich hier als Anwender und Organisator und nicht als Entwickler oder Hersteller von Produkten für die Kommunikationstechnik. In diesem Sinne bin auch ich Kunde der verschiedenen Produktlieferanten des Hauses und stehe in ähnlichen Situationen wie unsere externen Kunden, die ihr Kommunikationssystem (hier organisatorisch und technisch zu sehen) ausbauen und modernisieren wollen.

Zum besseren Verständnis der weiteren Ausführungen gebe ich Ihnen einen gerafften Überblick über den Aufbau unseres Unternehmens (Bild 1).

Sie sehen eine Gliederung in 6 Unternehmensbereiche: Den Unternehmensbereich Bauelemente, Daten- und Informationssysteme, Energietechnik, Installationstechnik, Kommunikationstechnik, Medizinische Technik und eine Vertriebsorganisation, die unsere Landesgesellschaften sowie Vertretungen im Ausland und die Zweigniederlassungen im Inland umfaßt.

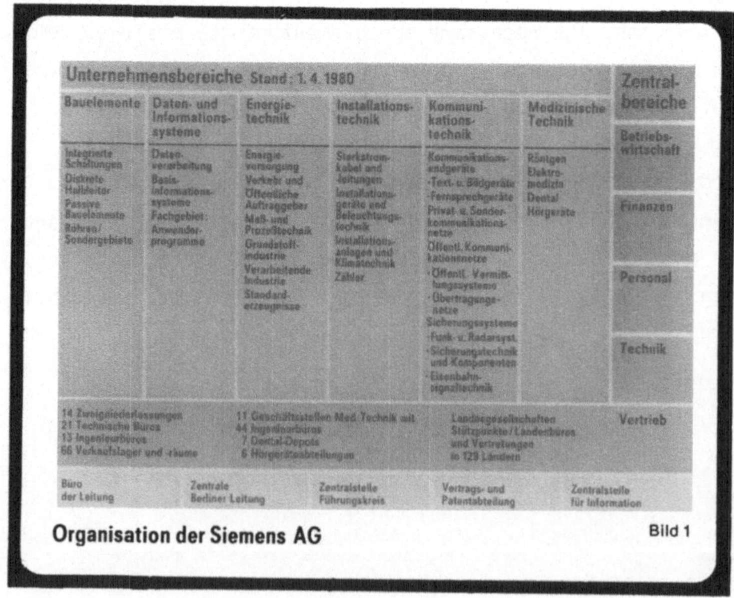

Organisation der Siemens AG

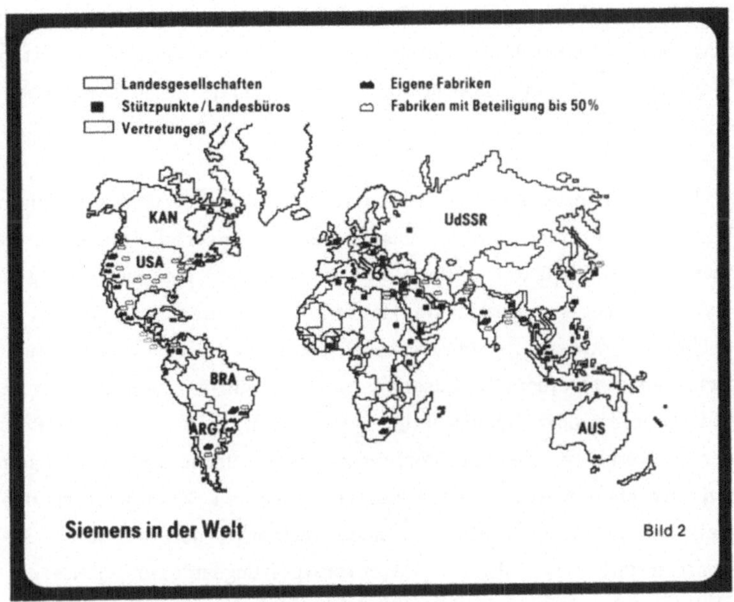

Siemens in der Welt
Bild 2

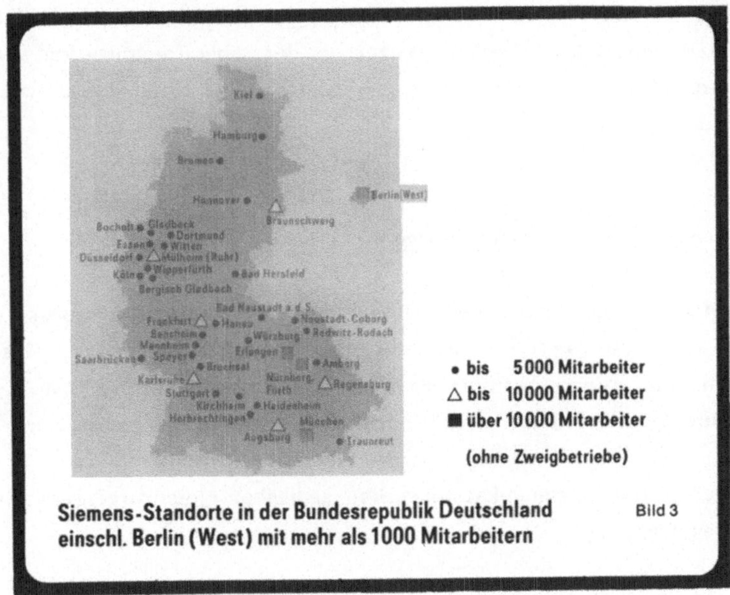

**Siemens-Standorte in der Bundesrepublik Deutschland
einschl. Berlin (West) mit mehr als 1000 Mitarbeitern**
Bild 3

Die Unternehmensbereiche werden von 5 Zentralbereichen unterstützt:

Betriebswirtschaft, Finanzen, Personal, Technik und Vertrieb (Bild 2, 3).

Von rund 335.000 Mitarbeitern sind 1/3 im Ausland und 2/3 im Inland tätig. Etwa die Hälfte aller Mitarbeiter haben ihre Arbeitsplätze im Büro- und Verwaltungsbereich.

Die Organisationsarbeit sowie die Unterstützung der Fachabteilungen bei der Rationalisierung der Arbeitsabläufe durch Bürotechnik und Datenverarbeitung werden von einer Vielzahl von Organisationsstellen durchgeführt, die in sämtlichen Teilbereichen angesiedelt sind.

Auf die Rationalisierungsmaßnahmen durch Automatisierung will ich nicht tiefgreifend eingehen. Der Stand der Automatisierung in unserem Hause ist hoch und hat in den letzen Jahren natürlich eine Reihe von personellen Auswirkungen gehabt. Dies betraf insbesondere die Stellen im Hause, die mit der Führung der verschiedensten Karteien befugt waren (Personalbüro, Dispositionsstellen, Einkauf etc.). Hier haben Dialogverfahren die personelle Zusammensetzung ganzer Dienststellen verändert. Konsequent dominieren dort höherwertige Arbeiten mit qualifiziertem Personal. Daß wir mit einer Vielzahl technischer Verfahren unsere Produktentwicklungen unterstützen, wird man mir ohne weitere Erläuterungen abnehmen. Zum Umfang der Automatisierungsarbeit sei an dieser Stelle nur noch vermerkt, daß wir allein im Inland ca. 80 Rechenzentren mit mehr als 125 Datenverarbeitungsanlagen mittlerer Größe betreiben. Die Anwender im Hause können dabei über ca. 5.000 Datensichtstationen eine Vielzahl ihrer Arbeiten im Dialog abwickeln. In den Organisationsabteilungen nehmen etwa 2.000 Mitarbeiter Verbesserungen in der Ablauforganisation vor, entwickeln Verfahren, pflegen Verfahren und produzieren neue Ideen zur Effizienzverbesserung des Gesamtsystems.

2. Problemlandschaft Kommunikation

Das weltweite Engagement des Hause im Entwicklungs-, Fertigungs- und Vertriebsbereich sowie die zentrale unternehmerische Verantwortung der Geschäftsbereiche erfordern sorgfältige Abstimmungsarbeit und damit eine intensive Kommunikation unter Nutzung aller wirtschaftlich einsetzbaren Kommunikationstechniken.

Einer weltweiten Kommunikation stellen sich aber gleich drei grundsätzliche Probleme in den Weg:

- Erstens der Zeitunterschied zwischen den Kontinenten, durch den sich teilweise Arbeitszeiten nicht mehr überschneiden und somit keine direkte Kommunikation zwischen den Partnern mehr möglich ist.

- Zweitens der unterschiedliche Stand der Industrialisierung in den Ländern, der sehr verschiedene Qualitäten der Kommunikationssysteme zur Folge hat.

- Drittens die politischen und verwaltungstechnischen Unterschiede zwischen den Ländern, die freizügige weltweite Kommunikation behindern.

In Anbetracht der immensen Zunahme internationaler Aktivitäten und der Wettbe-
werbe zwischen den Industrieländern wird jedes Unternehmen versuchen, sein inner-
betriebliches Kommunikationssystem effektiver zu gestalten. Es ist z.B. nicht mehr
tragbar, daß eine Ausschreibung aus Australien über ein größeres Projekt, das in
München bearbeitet werden muß, durch den herkömmlichen Transport für Hin- und
Rückweg nahezu 3 Wochen in Anspruch nimmt. In manchen Fällen bleibt somit kaum
Zeit, für die Bearbeitung dieses Vorgangs. Hier sind wir uns völlig einig mit den
Feststellungen der Kommission für den Ausbau des technischen Kommunikationssystems (KtK), die formuliert hat: "Der Wettbewerb auf den nationalen und internationalen Märkten richtet sich zunehmend auf die Frage, wer vor einer wirtschaftlichen Entscheidung schneller und exakter problembezogen mit treffenden Informationen versorgt ist, und wer mit aktiven Informationen schneller und wirkungsvoller
eingreift". Wettbewerb wird also "zunehmend Kommunikationswettbewerb."

In anderen Kapiteln kommt die Kommission zu Empfehlungen, die wir schon seit
vielen Jahren befolgen, wie z.B. den bedarfsgerechten Ausbau des weltweiten Telex-
Dienstes. Dieser Dienst ist derzeit der einzige allgemein zugängliche und international nutzbare für die schriftliche Kommunikation. Er bedient weltweit immerhin mehr
als 1,3 Mio Teilnehmer. Damit ist ein Großteil unserer Kunden und Zulieferanten
erreichbar. Den bedarfsgerechten Ausbau fördern wir innerbetrieblich durch die
Inbetriebnahme von Textnebenstellensystemen und dem breiten Einsatz von Bürofern-
schreibern.

Neuere Kommunikationsformen, die ihren Ursprung in der Datenfernverarbeitung haben, finden sich überwiegend in lokalen Installationen, die für den internationalen
Verkehr entweder Mietleitungen benötigen oder über das Fernsprechnetz nur festgelegte Teilnehmer erreichen. Solche lokalen Installationen sind auch in unserem Hause
sehr zahlreich. Außerdem fördern wir seit einigen Jahren auch die Bildkommunikation die ihren Schwerpunkt zur Zeit beim Fernkopierer hat. Hier werden wir bald
1.000 Geräte im Einsatz haben.

Um diese verschiedenen Kommunikationsformen in ihrem Einsatz für die internationale innerbetriebliche Kommunikation zu koordinieren und entsprechend den Bedürf-
nissen auszubauen, ist das Projekt SIEKOM in Angriff genommen worden.

3. Projekt SIEKOM

Die Projektarbeit konzentriert sich auf die Lösung technischer, organisatorischer
sowie Akzeptanz-Probleme. Technisch ist, wie wir alle wissen, fast alles machbar;
nur inwieweit es wirtschaftlich realisiert werden kann, ist oft erst nach genauer
Analyse der Verhältnisse zu beurteilen.

In Anbetracht der Vielfalt technischer Lösungen also besteht das generelle Problem für einen Organisator, die für die Situation geeigneten Mittel auszuwählen, die Akzeptanz beim Anwender zu erreichen und ein Höchstmaß an Effizienz - Schnelligkeit, einfache Bedienbarkeit, Wirtschaftlichkeit - in der Abwicklung sicherzustellen.

Unter Berücksichtigung der technischen und technologischen Innovationen sowie der möglichen organisatorischen Anpassunger und Lernfähigkeiten der Anwender sehen wir daher im Zeithorizont bis in die 90er Jahre folgende 3 Phasen (Bild 4).

Die vor uns liegende Zeit muß genutzt werden, um vorhandene Kommunikationsformen bedarfsgerecht weiter auszubauen, wie es auch in den Empfehlungen der KtK zum Ausdruck kommt.

Darüberhinaus müssen Pilotprojekte durchgeführt werden, in denen die Möglichkeiten von geplanten Diensten der Postverwaltungen berücksichtigt werden, die aber auch Privatsysteme (Standortlösungen) einbeziehen, wie z.B. Sprach- und Bildkonferenzen oder 'Electronic-Mail' im Büro.

Wir werden uns in dieser "Lernphase" damit auseinandersetzen, welche technischen Mittel der Kommunikation dem Fachprozeß (z.B. Auftragsabwicklung) am besten dienen, um später zu einer Strukturierung des technischen Kommunikationssystems zu kommen. Das heißt: Welche Mittel sind für welchen Zweck am effektivsten ?

Als Zwischenergebnis sehen wir teilintegrierte Kommunikationssysteme, die dem Anwender eine Reihe von Vereinfachungen bieten, z.B. bestimmte Kompatibilitäten, Reduzierung der Vielfalt von Geräten und Übertragungswegen. Auf der Basis der sich in diesem Zeitraum ergebenden Standards der Anwendung bestimmter Kommukationsformen, kann ein 'Integriertes Kommunikationssystem' schließlich Gestalt annehmen. In dem von uns als Konsolidierungsphase definierten Zeitraum müßten sich nach und nach die Benutzerwünsche und Benutzergewohnheiten in den technischen Leistungsmerkmalen der Einrichtungen ausgedrückt haben.

Akzeptanzbarrieren, wie sie im Beitrag von Herrn Professor Dr. Hauschildt*) aufgezeigt wurden und deren Auswirkungen wir täglich erleben, müßten dann mit entsprechenden Produkten, aufgeklärten Nutzern und durch die Arbeit qualifizierter Kommunikations-Organisatoren leichter überbrückt werden können. Die Konsolidierungsphase ist natürlich nicht das Ende der Entwicklung, sondern zeigt nur den Horizont, bis zu dem wir heute in groben Perspektiven vorausschauen können. Es ist z.B. nicht absehbar, welche "Verkehrsprobleme" im Bereich der Bild- und Text-Kommunikation entstehen, wenn entsprechende Bedürfnisse bei den Nutzern geweckt sind und ob sich das Kommunikationsverhalten durch Umwelteinflüsse wesentlich verändert.

---

*) Hauschildt, J., Barrieren für die Informationsnachfrage im Mensch-Maschine-Dialog. In: Telekommunikation für den Menschen. Hrsg.: E. Witte, Springer-Verlag, Berlin-Heidelberg-New York, 1980, S. 246-251.

Mit den folgenden Darstellungen möchte ich Ihnen nun die Hauptziele des Projektes SIEKOM aufzeigen.

Ausbau der innerbetrieblichen Kommunikationseinrichtungen für den Sprach-/Text-/Bild- und Datenverkehr unter Nutzung öffentlicher Dienste und privater Kommunikationssysteme

Allgemeines Ziel: Rationalisierung der Büroarbeit durch technische und organisatorische Verbesserung der Kommunikation

1. Kommunikationsbewußtsein verstärken
2. Notwendige Teilnehmerdichte (Erreichbarkeit) herbeiführen
3. Kommunikationsnetz ausbauen

PROJEKT SIEKOM

Bild 5

1. Kommunikationsbewußtsein verstärken

● Information der Führungskräfte aller Ebenen über Möglichkeiten und Effizienz technischer Mittel für die verschiedenen Kommunikationsformen

● Ausbildung von Organisatoren für die Kommunikation

● Entwicklung von Standort- und Landeskonzepten in enger Zusammenarbeit mit den zukünftigen Anwendern

● Einführung eines einfachen Berichterstattungssystems für den Informations- und Kommunikationsverkehr (Kontrollsystem)

● Schaffen des organisatorischen und finanziellen Rahmens für die Projektierung und Abwicklung überbereichlicher Kommunikation

PROJEKT SIEKOM

Bild 6

2. Notwendige Teilnehmerdichte (Erreichbarkeit) herbeiführen

- Intensivierung einzelner Kommunikationsformen (Bürofernschreiben, Fernkopieren, Ferntexten)

- Interklassenverkehr Daten-/Textkommunikation/Fax

- Neue Leistungsmerkmale beim Fernsprechen (z.B. Anrufumleitung, Konferenzschaltungen)

- Reduktion der Terminalvielfalt durch Einsatz neuer Technologien (Multifunktionsterminal)

Projekt SIEKOM

Bild 7

3. Kommunikationsnetz ausbauen

- Nebenstellensysteme für einzelne Kommunikationsformen verstärkt einsetzen

- Leitungsnetze gemeinsam für Sprache-/Text-/Bild- und Datenkommunikation nutzen

- Neue Leistungsmerkmale anhand von Pilotprojekten erproben

- Kompatibilitätsservice für unterschiedliche Endgeräte bereitstellen

- Erweiterte Vorkehrungen für Datenschutz und Datensicherung treffen

Projekt SIEKOM

Bild 8

4. Der Kommunikations-Organisator

Eine schrittweise Verbesserung des innerbetrieblichen Kommunikationssystems kann zwar zentral von einer qualifizierten Projektabteilung geplant werden, die organisatorische Detailarbeit muß aber in einem Großunternehmen von Fachleuten an Ort und Stelle ausgeführt werden. Ohne einen 'Kommunikations-Organisator' vor Ort ist die Einführung neuer Leistungsmerkmale für die verschiedenen Kommunikationsformen und die Einflußnahme auf das Kommunikationsverhalten der Anwender in der Breite nur schwer erreichbar. Mit einem Fachmann vor Ort können die Anwender ihre Erfahrungen austauschen, Änderungs- bzw. Verbesserungsvorschläge ausdiskutieren und qualifiziert an die Projektabteilung oder den Produktlieferanten herantragen. Zur raschen Förderung des hierfür nötigen Kommunikationsbewußtseins in allen Bereichen und auf allen Ebenen müssen dann auch genügend Kommunikations-Organisatoren zur Verfügung stehen.

An dieser Stelle möchte ich Ihnen gerafft das von uns angestrebte Anforderungsprofil für den Kommunikations-Organisator beschreiben. Dies ist kein abgeschlossener Vorgang und wie häufig in den Anfangsstadien der Dinge noch manchen Veränderungen ausgesetzt.

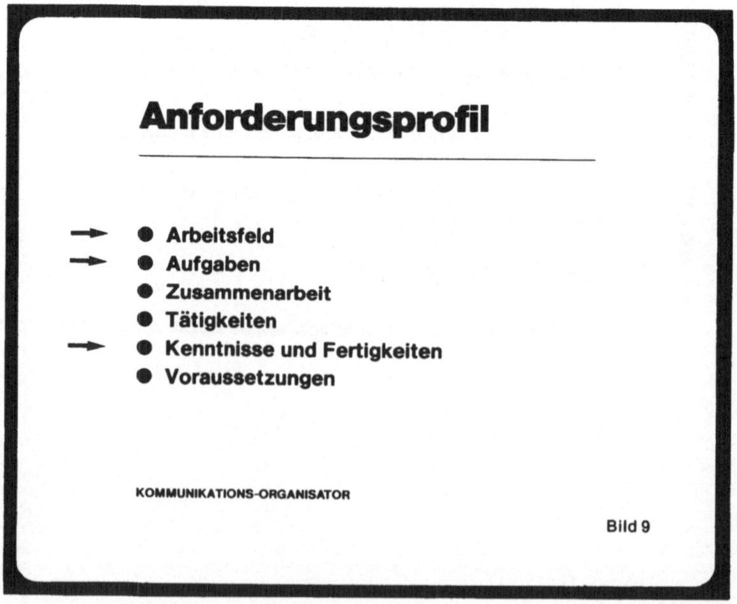

Bild 9

Arbeitsfeld Organisation und Einsatz von Kommunikationsmitteln wie:

- Fernsprechsysteme incl. Zusatzeinrichtungen
- Bildschirmtext, Videotext u. ä.
- Fernkopiertechnik
- Fernschreibtechnik (Tx, Ttx)
- Vermittlungssysteme (private)
- Übertragungseinrichtungen
- Endgeräte (Bildschirm, Hardcopy)
- Textsysteme mit Übertragungsanschluß
- Brief- und Rohrpostsysteme

Kommunikations-Organisator

Bild 10

Aufgaben

- Analysieren, Bewerten und Beschreiben von Problemen und Schwachstellen der Kommunikationsstruktur
- Erarbeiten wirtschaftlicher Sollkonzepte
- Abstimmen der Sollkonzepte mit betroffenen und weisungsbefugten Stellen incl. Realisierungsplan
- Realisieren des Sollkonzeptes
- Revolvierendes Controlling
- Betreuung und Beratung der Anwender und Bediener

Kommunikations-Organisator

Bild 11

Kenntnisse und Fertigkeiten	Auf folgenden Gebieten müssen Kenntnisse vorhanden sein oder erworben werden

- Betriebsorganisation
- Kommunikationstechnik
- Grundlagen der Betriebswirtschaft
- Textbe- und -verarbeitung
- Datenfernverarbeitung
- Planungs- und Analysemethodik
- Datenschutzgesetz, Benutzungsrecht für Kommunikationseinrichtungen
- Beschreibungs-, Strukturierungs- und Problemlösungstechniken
- Besprechungs- und Präsentationstechnik

Kommunikations-Organisator   Bild 12

5. Weitere Aspekte

Meine Damen und Herren, absichtlich habe ich mich bislang mit dem organisatorischen Aspekt des Problems beschäftigt, da sich die Fachpresse und die meisten Fachveranstaltungen überwiegend mit den technischen Neuerungen und deren Auswirkung auf die Menschen im allgemeinen auseinandersetzen. Diese Auseinandersetzung betrifft häufig auch noch zukünftige Kommunikationssysteme, deren Anwendung im betrieblichen Alltag noch nicht gegeben ist. Wahrscheinlich zeigt sich aus diesem Grund bei vielen Anwendern eine Verunsicherung über die sinnvolle weitere Vorgehensweise bei der Realisierung "Integrierter Kommunikationssysteme".

Für die traditionellen Kommunikationssysteme und die Datenverarbeitung sind in den letzten Jahrzehnten immense Investitionen getätigt worden, die in keinem Fall in ähnlichen Größenordnungen - aber in wesentlich kürzeren Zeitabständen - für neue Technologien aufgebracht werden können; und dies nicht zuletzt auch wegen der raschen technologischen Innovationen. Das organisatorische Vermögen, die neuen Möglichkeiten systematisch und kontinuierlich einem vernünftigen Nutzen zuzuführen, wird vermutlich den Ausschlag geben und nicht das Abwarten oder sofortige Eingehen auf Neues. Außerdem muß unterstrichen werden, daß lokale bzw. nationale Innovationen in der Breite erst erheblich später auf internationaler Ebene zur Verfügung stehen. Wenn auch Satellitenverbindungen in die Länder der Dritten Welt bestehen oder kurzfristig realisiert werden können, zeigen sich oft große Schwierigkeiten bei der Herstellung von Verbindungen zu den einzelnen Teilnehmern. Der Organisator, der weltweite Kommunikationswege für ein Großunternehmen erschließen soll, wird daher gleichrangig traditionelle wie neue Kommunikationstechniken berücksichtigen müssen.

Dies alles aber ist nicht möglich ohne die Anwender, die es "probieren" und mitmachen. Hier ist noch viel Aufklärungsarbeit zu leisten.

Die systematische technische Innovation der Kommunikationsformen in einem Unternehmen setzt also Partnerschaft mehrerer oder gar vieler Beteiligter voraus, die zum gleichen Zeitpunkt gemeinsame Lösungen benötigen, z.B. München – New York oder München – Sao Paulo. Für die überbereichliche und weltweite Kommunikation müssen die verschiedenen Inselsysteme direkt oder indirekt miteinander verknüpft werden. Dabei durchdringen sich Kommunikation und Informationsverarbeitung gegenseitig. Zudem müssen auch solche Büroarbeitsplätze einbezogen werden, die heute noch nicht mit technischen Hilfsmitteln ausgestattet sind.

Die Entscheidungsfindung für das richtige Maß der Investitionen auf diesem Gebiet, die sich bisher an der Effizienz und erreichbaren Wirtschaftlichkeit orientiert, erhält eine stärkere qualitative Betrachtungsebene. Neben der Wirtschaftlichkeit für Teilgebiete oder Nutzergruppen wird künftig der Nutzen des Gesamtsystems zu betrachten sein. Da im Anfangsstadium Vorleistungen zu erbringen sind, werden die qualitativen Vorteile, wie Schnelligkeit und Zeitgewinn im Vordergrund stehen. Die Bewertung dieser Kriterien ist schwierig. Hier müssen noch Grundlagen erarbeitet und Erfahrungen gesammelt werden.

DISKUSSION anschließend an den Vortrag von B. Czaputa

Frage: Gibt es theoretische Untersuchungen über die Bedeutung von
          Telekommunikation als Hilfe bei Entscheidungsprozessen?
          SIEKOM soll ja helfen, Entscheidungsprozesse besser und
          schneller abwickeln zu können.

Antwort: Natürlich belegen wir mit einer Vielzahl von Argumenten in
          dieser Richtung unser Vorgehen, das ist klar. Aber im Grun-
          de genommen kann man nur erreichen, daß genügend Informatio-
          nen rascher zur Verfügung stehen für einen Entscheidungspro-
          zeß. Aber den Inhalt, die Qualität des Entscheidungsprozes-
          ses kann man mit diesen Mitteln natürlich nicht beeinflussen.
          Es gibt da diverse theoretische Untersuchungen, die zum Aus-
          druck bringen, daß, selbst wenn 100% Information zur Verfü-
          gung stehen, der Mensch offensichtlich so geartet ist, daß
          er nur 10, 15% dieser Information auch wirklich nutzt. Eine
          Anhebung der Qualität der Information oder des Entscheindungs-
          prozesses wird durch eine bessere Kommunikationssystematik
          aus unserer Sicht im Moment nicht erreicht.

Frage: Die neuen Kommunikationsdienste bedingen sicherlich eine er-
          hebliche Änderung in den betrieblichen Abläufen eines großen
          Bürokomplexes. Welche Tendenzen bahnen sich an?

Antwort: Es gibt gemeinsame Überlegungen und eine Zusammenarbeit zwi-
          schen Mitarbeitern in den Organisationsabteilungen, die für
          die Datenverarbeitung zuständig sind, und denen, die für die
          Kommunikationsformen zuständig sind. Natürlich sehen wir fol-
          gendes: Wir befinden uns irgendwo in der dritten Automatisie-
          rungsphase, die es für uns nutzbar zu machen gilt. Nach der
          Entwicklung von Insel-Verfahren, nach der Entwicklung von
          Integrationsaspekten bei diesen Verfahren für den internen
          Ablauf befinden wir uns nun quasi in der Strukturierungsphase,
          die Dialogisierung hat seit Jahren erheblich zugenommen. Zur
          Dialogisierung kommt die verteilte Datenhaltung hinzu. Durch
          die Dialogisierung haben wir eigentlich den wesentlichen Ein-
          fluß auf das Arbeitsverhalten bekommen. Das gilt für die repe-
          titiven Aufgaben der Disponenten, für die Sachbearbeiter in
          der Fertigungsregelung, im Rechnungswesen, in der Kalkulation,
          im Einkauf. Dort haben sich die Dinge bereits schon in der

Form verändert, als wir für diese repetitiven Aufgaben erheblich weniger Mitarbeiter mehr beschäftigen, die dafür aber mehr Qualitätsarbeit machen. Ich denke an den Einkauf z.B., die Struktur des Einkaufs hat sich erheblich geändert, das Qualitätsniveau ist stark gestiegen. Natürlich ist durch die Elektronik auch eine neue Komponente für die Qualität der Mitarbeiter im Einkauf auf uns zugekommen, so daß wir zwar in der Summe fast genauso viele Mitarbeiter im Einkauf haben wie vorher, wenn man die einzelnen Fertigungsstätten betrachtet, aber die Zusammensetzung ist natürlich eine andere. Anderes Qualitätsniveau, andere Arbeit. Nun taucht ein neues Problem auf, das wir eben mit der Kommunikation lösen wollen. Sie könnten im Dialog da und dort sehr schnell disponieren, sich über einen Auftrag, einen Termin in Sekundenschnelle informieren. Wenn Sie aber dann Ihrem Partner eine schriftliche Nachricht geben wollen, schreiben Sie wieder einen ganz normalen Brief und das dauert wieder ein, zwei Tage, bis dieser Partner diese Bestätigung hat. Genau hier muß in diesem Prozeß der Abwicklung ein Einkaufs- oder Versandleiter in der Kommunikation, insbesondere wesentlich in der Textkommunikation, unterstützt werden, so daß also nach Möglichkeit gleich ein Briefentwurf entsteht für diese Bestätigung und er über die "Landstraßen-Systeme", die Lieferanten sind ja keine Großunternehmen in vielen Fällen, automatisch sozusagen über Telex dem Lieferanten oder dem Anfragenden zugeleitet wird. Also eine begleitende Unterstützung des Datenverarbeitungsprozesses, was man sicher auch mit anderen Kommunikationsformen machen kann, abhängig aber eben von den einzelnen Ländern, in denen man dieses Thema diskutiert. In dem Sinne gibt es also natürlich erhebliche Veränderungen in den betrieblichen Abläufen.

Frage: Sie haben erwähnt, daß Sie schon 1000 Faksimile-Geräte in Ihrer Organisation eingeführt haben. Sie sind einige Jahre schon mit der Schulung von Mitarbeitern und mit der Einführung gewisser organisatorischer Formen beschäftigt. Können Sie etwas sagen über die Akzeptanz und die Brauchbarkeit der Faksimile-Geräte und ob bei Ihnen schon gewisse Einsichten vorliegen bezüglich des Bürofernschreibers?

Antwort: Allgemein kann man sagen, jetzt greife ich mal das letzte Produkt Bürofernschreiber auf, daß in allen Großunternehmen,

Verwaltungsbereichen der Industrie die Verteilung oder Verbreitung des Bürofernschreibers offensichtlich eben noch nicht in Gang gekommen ist in der gewünschten Weise. Für ein Reisebüro oder für einen Rechtsanwalt ist ein Fernschreiber selbstverständlich, selbst wenn der tickt und laut ist. Wir haben mittlerweile erheblich leisere als die meisten Schreibmaschinen. Es scheint ein Organisationsproblem zu sein, von einer zentralen Bearbeitung in einem Schreibbüro abzukommen und wirklich einmal diese Information zum Sachbearbeiter über ein solches Gerät zu bringen. Die Überlegung und Diskussion dieses Themas bedeutet Innovation. Die Akzeptanz des Bürofernschreibers im Büro ist heute hervorragend. Wir verkaufen ca. 100000 Stück im Jahr. Wie gesagt, die Innovation aber auf dem Sektor in den Großverwaltungen ist ein Organisationsproblem. Das geht leider nicht so problemlos wie mit dem Taschenrechner. Hier müßten sich die Organisationsstellen in den Großfirmen mit dem Thema Kommunikation, mit diesem Aspekt der Datenkommunikation und deren Anforderungen auseinandersetzen, aber es kümmert sich zu wenig Personal darum.

Zum Thema Faksimile muß man sagen, daß wir natürlich erst in einem Entwicklungsprozeß uns befinden. Es gibt Faksimile-Geräte der ersten Generation. Wir kennen die Faksimile-Technik seit 20, 30 Jahren im Hause, trotzdem kam sie nicht vernünftig zum Einsatz. Erst als wir merkten, daß wir in Japan 120000 Geräte haben, oder in USA 150000 Geräte, plötzlich wachte man also auch da und dort auf und man beschäftigte sich mit diesem Produkt. Am Anfang hatte man das Gefühl, es gibt also eine Wettbewerbssituation zwischen Faksimile und Fernschreiben, z.B. das eine substituiert das andere. Wir haben mittlerweile längst bemerkt, daß es eine Reihe von Vorgängen gibt, die eben heute nicht mit einem Fernschreiber erledigt werden kann, sondern nur mit einem Faksimilegerät, weil dort Bilder, Skizzen, Grafiken übertragen werden. Für viele Personalabteilungen werden selbst Personalvorgänge übertragen, die sehr schnell und vertraulich von einem Ort zum anderen vermittelt werden müssen. Als Fernschreibnachrichten könnte man sie eventuell abhorchen, das ist beim Faksimile zunächst mal nicht gegeben. Nun gibt es für mich in den nächsten Stufen Verbesserungen bei den

Fernkopierern. Aber auch das ist ein Organisationsproblem.
Trotzdem bleibt heute ein Problem der Akzeptanz, das ist
nämlich das Problem der Gerätevielfalt und das Problem rein
räumlich in den Büros die Geräte unterzubringen. Das ist
wirklich ein Problem und deswegen sehe ich in der Zukunft
dieses Multifunktionsterminal auf uns zukommen.
Die Bürofernschreiber mit weiteren Leistungsmerkmalen, das
sind also alles für uns evolutionäre Schritte und Gewohn-
heitsschritte, eine Vorstufe des automatisierten Büros.

Frage: Könnten Sie als Fachmann für innerbetriebliche Kommunikation
eine kleine Bewertung versuchen über die neuen Dienste, be-
ginnend mit langsamer Textkommunikation á la Telex, schneller
Textkommunikation â la Teletex, langsamer Bildkommunikation
á la Faksimile, schneller Textbildkommunikation und vielleicht
Bewegbildkommunikation. Was ist aus Ihrer Sicht der Nutzen
dieser neuen Dienste für die innerbetriebliche Organisation?

Antwort:Wir haben uns längst daran gewöhnt, daß wir ein breit gefächer-
tes Angebot im Verkehr haben. Es gibt den Schiffsverkehr, es
gibt den Flugverkehr und es gibt Eisenbahnen und es gibt
schnelle und langsame Autos und es gibt auch noch ein paar
Postkutschen. Ich bin der Meinung, daß wir auch im Bereich
der Kommunikation diese Vielfalt bekommen werden. Das wird
leider überlagert durch die Tatsache, daß der Industriali-
sierungsstand in den Ländern sehr unterschiedlich ist. Sie
müssen also auch Teilnehmer in Entwicklungsländern erreichen können, die sich nicht Protokollstandards anschließen.
Also es wird darauf ankommen, daß Sie für die verschiedensten Aufgaben und für die verschiedensten regionalen Ober-
bereiche in Kommunikationsarten Lösungen suchen.

Was also vielleicht einesteils Teletext und, wir hoffen
sehr,auch Bildschirmtext bietet, muß man orientieren an
den Leistungsmerkmalen, die man eben innerbetrieblich
braucht. Wenn Sie eine Informationsbank für Bildschirmtext einrichten, die ausschließlich einem bestimmten
Zweck dient, werden Sie nicht umhin kommen, Bildschirmtext-
terminals einzusetzen. Nun wird es darauf ankommen, was in
diese Datenbank alles hineingebaut wird. Möglicherweise
gibt es eine Reihe von Electronic Mail Komponenten, die

dann ein anderes Produkt, das sich auch mit Electronic Mail
beschäftigt hat und vielleicht nicht den Komfort bietet,
langsam zum Aussterben bringt. Aber ich glaube, daß es
hier gegenseitige Einflüsse geben wird. Aber das ist
das von mir geschilderte Problem. Wir werden in den nächsten Jahren mit diesen Techniken Erfahrungen sammeln müssen, wir werden sie an den Automatisierungsprozessen der
Datenverarbeitung ein wenig anbinden, wir werden diese
Kommunikation unterstützen und dann wird sich, wir sind
sehr überzeugt, in der Mitte der 80er Jahre herauskristal-
lisieren, welche Kommunikationsarten lokal, überregional
und international für uns relevant sind.

Stand und zukünftige Planungen
für neue Telekommunikationsdienste in Europa

H. Kunze
Darmstadt

Zusammenfassung

Grundlage der Nachrichtenkommunikation sind Basisnetze, die Grundleitungen für die unterschiedlichsten Anwendungen bereitstellen.
Mit reinen Transportfunktionen im Netz sind jedoch "value added
services" nicht darstellbar. Im ersten Teil werden heutige und
zukünftige Netzkomponenten einschließlich der Satelliten-Technik
diskutiert. Hierbei werden auch aktive Netzfunktionen (Prozedur/
Format-Wandlungen, Netzunterstützung) behandelt. Im zweiten Teil
werden die vom CCITT definierten Benutzerklassen vorgestellt.
Hierbei zeigt sich, daß praktisch alle allgemeinen Anwendungsfäl-
le mit den notwendigen Schnittstellen für heutige und zukünftige
Datenübermittlung einschließlich der notwendigen Netzübergänge
bereits spezifiziert sind. Die Realisierung in bereits bekannten
Netzen ist dabei möglich. Die Notwendigkeit besonderer value
added networks ist nicht zu erkennen. Im dritten Teil werden
einige Dienste beschrieben, die ebenfalls von normierten Daten-
übertragungsverfahren Gebrauch machen. Zusätzlich werden jedoch
bestimmte Zeichenvorräte und Prozeduren definiert, die dann auch
bestimmend für die Gestaltung der Endeinrichtungen sind. Diese
Dienste stehen alle in Zusammenhang mit dem Oberbegriff
"electronic mail".

Summary

The fundamentals of data communications are base networks which
provide the groundwork links for the different applications. The
mere transport service, however, is not sufficient for value
added services.

The first part of the lecture discusses present and future network
components including satellite technology. Active network functions
like protocol/format conversions and network control support are
explained.

The second part presents the CCITT defined classes of users. It is
pointed out that there have already been specified communication
interfaces for practically all general applications including
internetworking. These interfaces may be implemented in the
existing networks. Therefore it seems not to be necessary to
have special "value added networks".

The third part describes some services which are based on
standardized data transfer technologies. They may be
specified by special terminal equipment which make use of
additionally defined character codes and procedures.
All these services are related to "electronic mail".

Einleitung

Grundlage für alle Dienste sind optimal ausgebaute Netze. Daher
hatte der Münchner Kreis als Veranstalter vorgeschlagen, die Darstellung der Basisnetze und deren technologischen Hintergrund mit
Übertragungs- und Vermittlungstechnik für

Fernsprechen
Datenübermittlung und
Breitbandkommunikation

an den Anfang zu stellen. Das ist auch durch die andere Situation
in Europa begründet. Alle Vorredner haben nämlich ausführlich über
value added networks (VAN) vorgetragen, ohne sich all zu viele Ge-
danken über die Bereitstellung der hierfür notwendigen Grundleitungen zu machen. Die europäischen Verwaltungen sind verpflichtet,
Grundleitungen und Dienstleitungen unterschiedlichster Art zur
Verfügung zu stellen.

Der Vortrag wird daher aus drei Teilen bestehen:

Netze

Allgemeine Datenübertragung, die allen Anwendungen aufgrund
international festgelegter Benutzerklassen und Leistungsmerkmale zur Verfügung steht.

Neue Dienste mit speziellen Bezeichnungen wie Teletex u.a.

1. Netze

Heutige Netze können grob in drei Bereiche aufgeteilt werden

Fernsprechnetze

analog o,3 - 3,4 KHz
einschließlich Datenübertragung im Wählnetz und auf besonderen
Leitungen (HfD im Bereich der DBP)

digital 64 KBit/s

Datennetze

digital $\leqq$ 48 KBit/s

Breitbandnetze

  im Funkbereich, Videosignal 5 MHz (analog)

  im Kabelbereich, Videosignal 5 MHz (analog)
  bzw. digital ⟩ 64 KBit/s als Einzelkanal

Aus dieser Aufzählung ist auch die Strategie der bisherigen Vorgehensweise zu erkennen. Es werden nämlich Basisbandkanäle (0,3 -
3 KHz, 64 KBit/s, 5 MHz) zur Verfügung gestellt, in die sich dann
die unterschiedlichsten "Dienste" einfügen können. Dieser Weg muß
auch zukünftig eingehalten werden, weil es sicherlich nicht wirtschaftlich ist, "Dienste" mit <u>für diesen Dienst</u> z. B. optimalen
Merkmalen zu definieren und dann zu fordern, daß das Netz diese
nicht standardisierten Geschwindigkeiten übertragen kann.

Für zukünftige digitale Netze wird der <u>Standardkanal</u> 64 KBit/s
haben. Daß dieser vom CCITT vorgeschlagene Weg richtig ist, zeigt
auch die heutige Situation im Fernsprechnetz. Im heutigen Basiskanal von 0,3 - 3,4 KHz findet in sehr großem Umfang auch Datenübertragung bis zu 9,6 KBit/s statt, also eine Anwendungsart, die
ursprünglich für dieses Netz gar nicht vorgesehen war.

Neben dem Fernsprechnetz gibt es in Europa bereits in großem Umfang eigenständige digitale Datennetze mit Geschwindigkeiten
≦ 9,6 KBit/s. Demnächst wird die Geschwindigkeit 48 KBit/s zum
allgemein angebotenen Standard gehören.

Den dritten Bereich bilden die Breitbandnetze. Der Normkanal ist
heute der Videokanal mit 5 MHz Bandbreite, weil noch kein anderer
Bedarf vorhanden war. Das gilt für die Funk- und Kabelnetze einschließlich Kabelfernsehen (KTV). Kabelfernsehen als eine Untermenge der Breitbandkommunikation soll hier nicht weiter behandelt
werden. Zukünftige Netze werden aber auch Signale ⟩ 64 KBit/s
übertragen. Die Übermittlung von <u>digitalen</u> Signalen ⟩64 KBit/s
wird als Breitbandkommunikation bezeichnet. Hierunter fallen
dann auch heutige Videosignale in digitaler Form, aber auch
Signale, die z. B. als Untermenge den digitalisierten Fernsprech-
kanal 64 KBit/s enthalten.

Der erste Schritt muß jedoch die Umstellung des heutigen Fern-
sprechnetzes auf 64 KBit/s sein. Wenn nämlich Netze mit <u>trans-
parentem</u> 64 KBit/s-Kanal zur Verfügung stehen, können sehr viele
Teilnehmerwünsche bereits realisiert werden.

Die Gliederung der Netze mit Wählvermittlung zeigt das Bild 1. In
dieser Gliederung sind die Breitbandnetze nach vorheriger Definition nicht enthalten. Endziel der Entwicklung wird es sein, alle
Netze in einem integrated services digital network (ISDN) zusammen-
zufassen. Bei digitalen Fernsprechnetzen werden international
Studien durchgeführt mit dem Ziel, die Anschlußleitung mit einer
Bitrate (64 + n) KBit/s zu betreiben, wobei $n \ll 64$ sein soll. Mit
n KBit/s könnten dann neben dem Fernsprechen zusätzliche Dienstleistungen auf einer Doppelader übertragen werden. Dieses Verfahren
wird in erster Linie für die digitale Ausnutzung von <u>vorhandenen</u>
Kupferadern Bedeutung erlangen. Im Bereich der Text- und Datenver-
mittlung gewinnt die Teilstreckenvermittlung in Form der Paketvermittlung eine größere Bedeutung. Euronet ist in Betrieb. Es gibt
also nicht nur Tymnet und Telenet. Es muß auch darauf hingewiesen
werden, daß europäische Teilnehmer bereits seit längerem die Möglichkeit haben, über PTT-Netzknoten mit diesen Netzen zu verkehren.

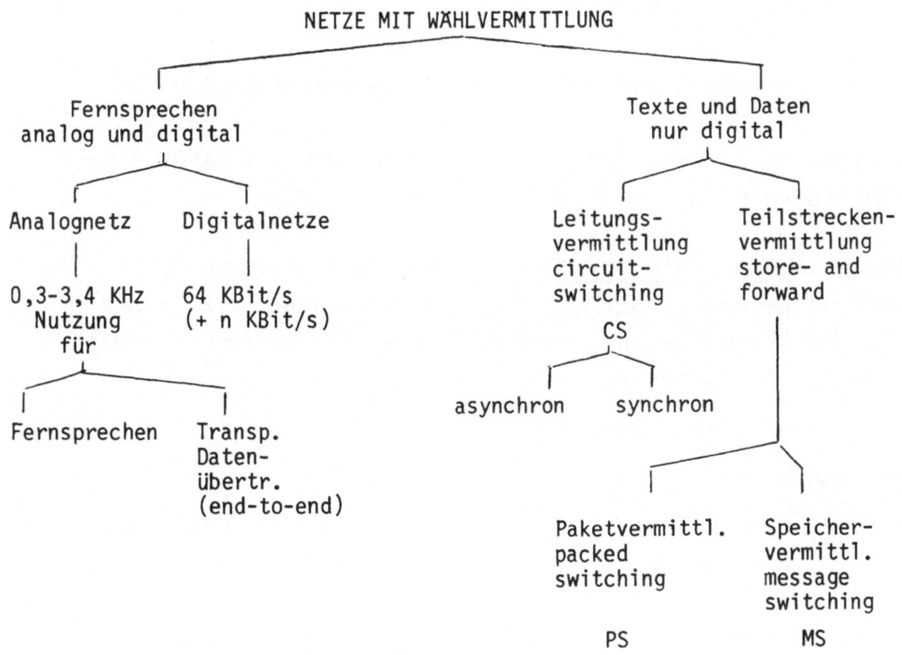

Bild 1

1.1. Merkmale der Fernsprechnetze

Die Multiplexsysteme der <u>analogen Übertragungstechnik</u> sind vom CCITT standardisiert worden und zeichnen sich durch folgende Merkmale aus:

Frequenzmultiplex von 12 Kanälen (Primärgruppe, 48 KHz) bis 10.800 Kanäle (60 MHz-System) bzw. 14.400 Kanäle (80 MHz) für zukünftige Anwendungen.
Hierbei Ausnutzung auch für Daten durch MODEM-Einsatz

Basisband bis 9,6 KBit/s, d. h. ca. 4 $\frac{\text{Bit/s}}{\text{Hz}}$

Primärgruppe (48 KHz) → 64 KBit/s

Tertiärgruppe (1,3 MHz) → 2 MBit/s

Übertragungsmedium

Cu-Kabel, symmetrisch

Cu-Kabel, koaxial

Richtfunk <10 GHz

Lichtwellenleiter (für Versuche)

Für zukünftige Anwendung als Schnittstelle zwischen Analog- und Digitalnetzen muß noch der Transmultiplexer entwickelt werden,um <u>frequenzoptimal</u> z. B. zwei <u>digitale</u> PCM 30-Systeme in einer <u>analogen</u> Sekundärgruppe (mit 60 Kanälen) zu übertragen.

Die Situation der <u>analogen Vermittlungstechnik</u> ist nicht durch CCITT-Standards gekennzeichnet. Lediglich für Koppelnetzeigenschaften in übertragungs- und verkehrstheoretischer Hinsicht gibt es Empfehlungen. An Systemvarianten finden wir

Direktwahltechnik (step-by-step)

Crossbartechnik

Rechnersteuerung mit gespeichertem Programm (stored program control, SPC) und unterschiedlichsten Koppelnetzen

Bei SPC-Systemen werden zahlreiche Leistungsmerkmale mit Netzunterstützung angeboten, z. B.

automatischer Weckdienst

Anrufumleitung

Ruhe vor dem Telefon u. a. m.

Ca. 80 Leistungsmerkmale sind von der CEPT empfohlen und damit
standardisiert worden. Aber auch heutige Netze können Merkmale
anbieten, die ursprünglich nicht vorgesehen waren, wie z. B. im
Bereich der DBP die Fernsprechkonferenz. Alle diese Merkmale
können nach der in den USA üblichen Definition als value
added services bezeichnet werden.

Diese bereits empfohlenen Merkmale müssen in gleicher Form zukünftig auch von Digitalnetzen und damit den digitalen Vermitt-
lungen angeboten werden. Bei sehr vielen Merkmalen ist <u>SPC das
Entscheidende</u>, nicht analog oder digital.

Die <u>Kostensituation für Analog- und Digitaltechnik</u> zeigt Bild 2.
Rein analoge Netze zu 100 % angesetzt, ergab eine Studie des BPO:

analoge ÜT/digitale VT → 90 % der Kosten

digitale ÜT/analoge VT → 80 % der Kosten

digitale ÜT + VT   → 45 % der Kosten

Diese Ergebnisse wurden auch durch eigene Studien der DBP und
anderer Verwaltungen voll bestätigt. Das Ziel der britischen Verwaltung ist ebenfalls aus Bild 2 zu ersehen. Alle europäischen
Verwaltungen bemühen sich zur Zeit, den Einsatz der digitalen ÜT
zu forcieren und möglichst frühzeitig mit der digitalen VT nach-
zuziehen. In einigen Ländern werden bereits in größerem Umfang
digitale Transit-Vermittlungen für das Fernnetz projektiert. Vor
1990 wird es jedoch kaum möglich sein, einigermaßen flächendeckende
digitale Fernsprechnetze zu realisieren.

Die <u>Übertragungstechnik</u> des Digitalnetzes mit dem Basiskanal
64 KBit/s wurde durch CCITT weitgehend standardisiert:

   Zeitmultiplex von PCM 30 (2 MBit/s) bis PCM 7 680 (565 MBit/s)
   bzw. später auch ca. 1,2 GBit/s

Die Ausnutzung dieser Multiplexsysteme soll durch Sprache und
Daten aller Geschwindigkeitsklassen sowie Breitbandkommunikation
erfolgen.

   Übertragungsmedium

      Cu-Kabel, symm. vorwiegend für PCM 30

      Cu-Kabel, koax. bis PCM 1920
      (140 MBit/s) bzw. als Übergang auch 565 MBit/s

COMPORATIVE COST OF NATIONAL TOLL NETWORK - Studie des BPO

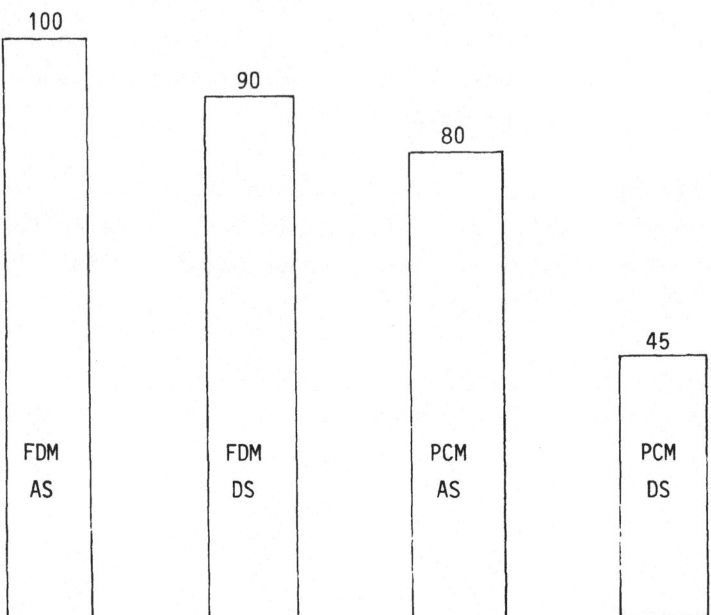

FDM : Frequenzmult.   AS : analoge Verm. (NF)

PCM : Zeitmultiplex   DS : digitale Vermittlg.

Die Angaben gelten für das Netz bei jeweils 100%igem Ausbau mit diesen Kombinationen.

Ziel ist es, bei Einführung der DS (in ca. 2 Jahren?) bereits in großem Umfang auf PCM-umgestellte Umgebung zu treffen.

Daher Umrüstung:

8 MBit/s auf vorhandenen Tf-Kabeln

11 GHz-Richtfunk

140 MBit/s auf Koaxial-Kabeln

Bild 2

Richtfunk bis 140 MBit/s im Bereich 2 - 15 GHz

Lichtwellenleiter für alle Bereiche der Breitbandkommunikation.

Die Aufstellung zeigt, daß das vorhandene Cu-Kabel-Netz für die nahe Zukunft noch eine große Bedeutung haben wird. Die Bedeutung des Richtfunks für den Aufbau digitaler Netze wird jedoch stark zunehmen. Das zukünftig bedeutendste Übertragungsmedium wird jedoch der Lichtwellenleiter ("Glasfaser") werden. Zahlreiche Versuchssysteme werden bereits für den Regeleinsatz in Fernsprechnetzen betrieben.

Für die digitale Vermittlungstechnik wurden Koppelnetzeigenschaften und die Programmiersprache CHILL vom CCITT standardisiert. Wesentliche Merkmale sind

SPC

weitgehend modularer Aufbau mit dezentralem Mikroprozessor-Einsatz

Durchgangsvermittlungen (Transit) werden in einigen europ. Staaten ab 1981 verfügbar sein mit der 2 MBit/s-Schnittstelle bzw. zukünftig auch 8 MBit/s.

Teilnehmer-Vermittlungen mit analogen Anschlußleitungen sind ebenfalls ab 1981 verfügbar.

Die für den endgültigen Einsatz benötigten Teilnehmer-Vermittlungen mit digitalen Anschlußleitungen sind noch in Diskussion. Das Problem ist hierbei die Betriebsweise auf der digital ausgenutzten Cu-Anschlußleitung, die zumindest in der Übergangsphase wegen ihrer großen Verbreitung genutzt werden muß. Es bietet sich hierbei an, die Cu-Doppelader mit (64 +n) KBit/s zu nutzen. Die wesentlichen Fakten zeigt Bild 3. Die für zukünftige "Dienste" (z. B. Teletex) benötigte Bitrate von n KBit/s wird in der Endstelle dem "Fernsprechen" mit 64 KBit/s zugesetzt und vor der Fernsprech-Digitalvermittlung wieder abgezweigt und den bereits vorhandenen anderen Netzen, u. U. über Multiplexer, zugeführt. Verschiedene mögliche Betriebsweisen werden z. Z. durch einen ständigen Kern der CEPT studiert. Vor zwei Jahren (einschl. Versuche) ist kaum mit einer Empfehlung zu rechnen. Man sollte unbedingt diese Standardisierung abwarten, um Fehlentwicklungen zu vermeiden.

DIGITALE ANSCHLUSS - LEITUNG
AUF CU - DOPPELADERN

NOTWENDIG ALS VORSTUFE ZU EINEM DIENSTINTEGRIERTEN NETZ (ISDN)

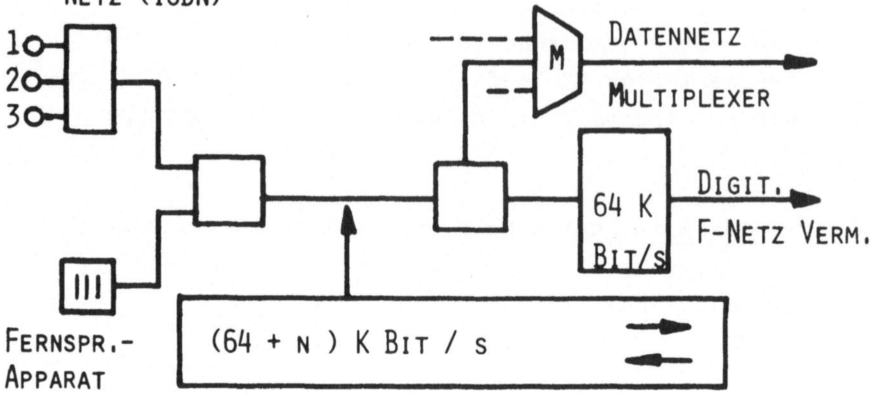

1,2,3: TRANSPARENTE EIN/ AUSGÄNGE FÜR $\leq$ 9,6 K BIT/S (NEUE DIENSTE).

OFFEN IST N = 8,12,16 . . . ?

BETRIEB AUF DER LEITUNG:

- VIERDRAHTVERF. MIT KLEINKONZENTRATOR
- ZEITGETRENNTLAGEVERFAHREN (BURST)
- FREQUENZGETRENNTLAGEVERFAHREN
- ZWEITDRAHTGLEICHLAGEVERFAHREN (GABEL)

STUDIUM DURCH STÄND. KERN DER CEPT

Bild 3

In einem vorhergehenden Vortrag wurde gesagt, die Standardisierung behindere Innovationen. Sie solle daher sehr spät erfolgen. Dieser Auffassung kann nur sehr bedingt gefolgt werden. Eine zu späte Standardisierung bzw. ein Aufbau vor der Standardisierung macht es erfahrungsgemäß fast unmöglich, z. B. verschiedene Netze miteinander zu verbinden. Mir ist es in diesem Zusammenhang auch nicht klar, ob z. B. der Zustand, daß ein Teilnehmer von X - TEN nicht

mit einem Teilnehmer von Telenet oder umgekehrt verbunden werden
kann, langfristig durchgestanden werden kann. In Europa bemühen
sich alle Verwaltungen, diese Diskrepanz durch striktes Einhalten
aller internationalen Empfehlungen zu vermeiden. Um nun Kunden-
wünschen und technologischen Entwicklungen zügig folgen zu können,
ohne dabei verlorenen Aufwand zu haben, ist frühzeitige Standardisierung notwendig.

In mehreren Vorträgen wurde auch auf den Iststand der Glasfasertechnik
in den USA eingegangen. Selbstverständlich wird auf diesem Sektor
auch in Europa viel getan. Für die am häufigsten verwendete Gradientenfaser zeigt Bild 4 die wesentlichen Eigenschaften. Z. Z. wird der
0,8 μm-Bereich genutzt. Leider ist die Dämpfung in diesem Bereich
um mehr als den Faktor 4 größer gegenüber den Bereichen 1,2 bzw.
1,6 μm. Die für diese Bereiche notwendigen technologischen Voraus-

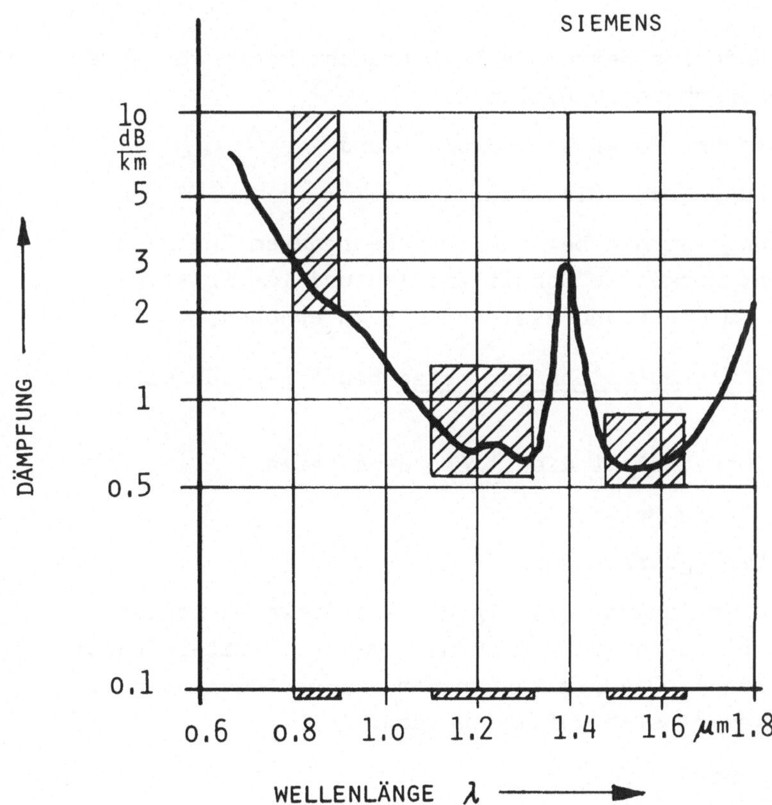

SPEKTRALE DÄMPFUNGSKURVE EINER GRADIENTENFASER

Bild 4

setzungen (optische Sender und Empfänger) sind jedoch für einen
betrieblichen Einsatz noch nicht vorhanden. Trotzdem wird überall
für kürzere Reichweiten mit der Nutzung der 0,8 μm-Technik be-
gonnen. Für das Weitverkehrsnetz sollte jedoch wegen der dann sehr
großen "verstärkerlosen Reichweite" auf die Betriebsreife der
1,2/1,6 μm-Technik gewartet werden.

Die französische Verwaltung hat beschlossen, das Ortsnetz
Biarritz in größerem Umfang nach verschiedenen Verfahren für
Versuchszwecke auszubauen. Ähnliche Projekte im Bereich der
DBP laufen im ON Berlin (West).

1.2. Datennetze in Europa

Die Datenübermittlung kann von drei Standpunkten aus betrachtet
werden: Anwender, Netzbetreiber, Hersteller. Dabei ist die Inter-
essenlage durchaus unterschiedlich.

Der <u>Anwender</u> als Kunde des Netzbetreibers braucht kostengünstige,
zuverlässige und überschaubare System für

    Datenfernverarbeitung (Datenübertragung) und

    Bürorationalisierung (neue Dienste).

Die technische Lösung ist hierbei von untergeordnetem Interesse.
Der Anwender wünscht normierte Schnittstellen und langfristig
stabile Verfahren, um die Planungssicherheit zu erhöhen.

Die europäischen PTT als <u>Betreiber der Netze</u> benötigen aus wirtschaftlichen Gründen

    universell anwendbare, <u>nicht</u> dienstgebundene Netze

    kostengünstig wartbare Netze

    Netze mit hoher Verfügbarkeit und

    Strukturen mit hoher Anpassungsfähigkeit an neue Technologien
und Konzepte (z. B. Austausch von Kern- gegen Halbleiter-Speicher
<u>ohne</u> Konzeptänderung). Dadurch kann neuen Innovationen schnell
und ohne "Reibungsverluste" gefolgt werden.

Der <u>Hersteller</u> braucht

    weltweit vertreibbare Produkte und

    wirtschaftliche, also nicht zu kurze Innovationszyklen.

Es ist selbstverständlich. daß auch der Netzbetreiber zu kurze
Innovationszyklen nicht realisieren kann.

Bei der Betrachtung des Iststandes der Datenübertragung in
Europa wird jetzt immer wieder CCITT erwähnt werden. Es sollte
auch nicht zugelassen werden, daß ein europäisches Land CCITT-
Standards nicht beachtet oder nicht zulässige Ergänzungen
(options) einführt. Im selben Augenblick ist dann die Netz-
kompatibilität nicht mehr vorhanden. Die Bilder 5 und 6 zeigen
den augenblicklichen Zustand in Europa. Vorhanden sind leitungsvermittelte asynchrone Netze (cs), leitungsvermittelte synchrone
Netze sowie paketvermittelte Netze (PS) mit dem von der EG for-
cierten Euronet. Die Merkmale dieser Netze sind durch 11 Benutzerklassen nach CCITT eindeutig definiert. Durch diese Benutzerklassen
werden bestimmte Datenübertragungsgeschwindigkeiten und andere
Merkmale vorgegeben. Aus Bild 6 ist zu ersehen, daß der Istzustand
in Europa nicht einheitlich ist. Es gibt auch Länder, die überhaupt keine leitungsvermittelten Netze einführen werden. Es wird
daher zwischen diesen beiden Netztypen Schnittstellen geben müssen,
um Übergänge von dem einen Netztyp in den anderen Netztyp zu
schaffen. Die Verbreitung von Netzen mit Nachrichten-Speicherver-
mittlung (message switching, MS) und den Stand auf dem Gebiet der
digitalen festen Verbindungen zeigt das Bild 7. MS-Netze bzw. MS-
Merkmale werden nur in F, GB, I und E angeboten. Es handelt sich
hier um dieselben Länder, die relativ frühzeitig auch die Paket-
vermittlungstechnik eingeführt haben (hierzu auch Bild 1). Zahlreiche PTTn bieten ihren Kunden festgeschaltete Verbindungen an,
z. B. im Bereich der DBP den Hauptanschluß für Direktruf (HfD)
und andere Leitungen. Auch hierfür gibt es festgelegte Benutzerklassen, durch die erst die grenzüberschreitenden Leitungen möglich wurden. Durch z. B. die Benutzerklasse 7 wird die Geschwindigkeit 48 KBit/s definiert.

Das Bild 8 zeigt die EurOnet-Konfiguration. Das Netz ist seit
März 80 voll in Betrieb. Sämtliche Staaten der EG haben Zugriff
zu diesem Netz. Entsprechend CCITT-Empfehlung werden die Benutzerklassen 8 - 10 (max. 9,6 KBit/s) und der Zugang aus anderen
Netzen bzw. umgekehrt (PAD-Funktion) angeboten. Zweck ist in
erster Linie, die wissenschaftlichen Datenbanken Europas optimal
miteinander zu verbinden. Das Netz wird von den PTTn betrieben
und ist daher ein öffentliches Netz. Bild 9 zeigt die Größen-
ordnung einiger Gebührenelemente dieses Netzes. Die nähere Betrachtung der Gebührenelemente zeigt, daß wegen des entfernungs-
unabhängigen Tarifs die Chancengleichheit für alle Teilnehmer

| | CS | | CS | | PS | | PS |
| | Leitungs-vermittelt asynchron | | Leitungs-vermittelt synchron | | Paket-vermittelt | | Euronet |
Land	1980	1983	1980	1983	1980	1983	1980 **
Belgien	−	*	−	*	−	*	* 1980
Bundesrep.D.	*	*	*	*	*	*	* 1980
Dänemark	*	*	*	*	−	○	* 1980
Finnland	*	*	*	*	−	○	− −
Frankreich	−	−	−	−	*	*	* 1980
Großbritannien	−	○	−	−	*	*	* 1980
Irland	−	○	−	−	−	*	* 1980
Italien	*	*	−	*	−	*	* 1980
Luxemburg	−	−	−	−	○	*	* −
Niederlande	−	−	−	−	*	*	* 1981
Norwegen	*	*	*	*	○	*	○ 1981
Österreich	*	*	−	*	−	○	− 1981
Portugal	−	○	−	○	−	○	− −
Schweden	−	−	*	*	○	*	○ 1981
Schweiz	*	*	−	○	○	*	* 1981
Spanien	−	−	−	○	*	*	○ 1981
Benutzer-Klassen nach CCITT	1 - 2 asynchron		3 - 7 synchron		8 - 11 und PAD		8 - 10 und PAD

* verfügbar

○ Einführung erwogen, in Untersuchung

− nicht verfügbar, bisher nicht vorgesehen

** Jahr für voraussichtliche Verknüpfung zwischen
   Euronet und nationalen Datennetzen

PLÄNE FÜR DIE EINFÜHRUNG VON ÖFFENTLICHEN

DATENNETZEN IN CEPT-LÄNDERN

CS : circuit switching
PS : packet switching
PAD: packet assembly/disassembly (CCITT X 3)

Bild 5

Zugang zu Überseenetzen,
auch Telenet, Tymnet
möglich
DATENNETZE IN WESTEUROPA STAND 1980
PAKETVERMITTLUNG(PS)
LEITUNGSVERMITTLUNG(CS)
DATENNETZE IN WESTEUROPA STAND 1983
PAKETVERMITTLUNG(PS)
LEITUNGSVERMITTLUNG(CS)

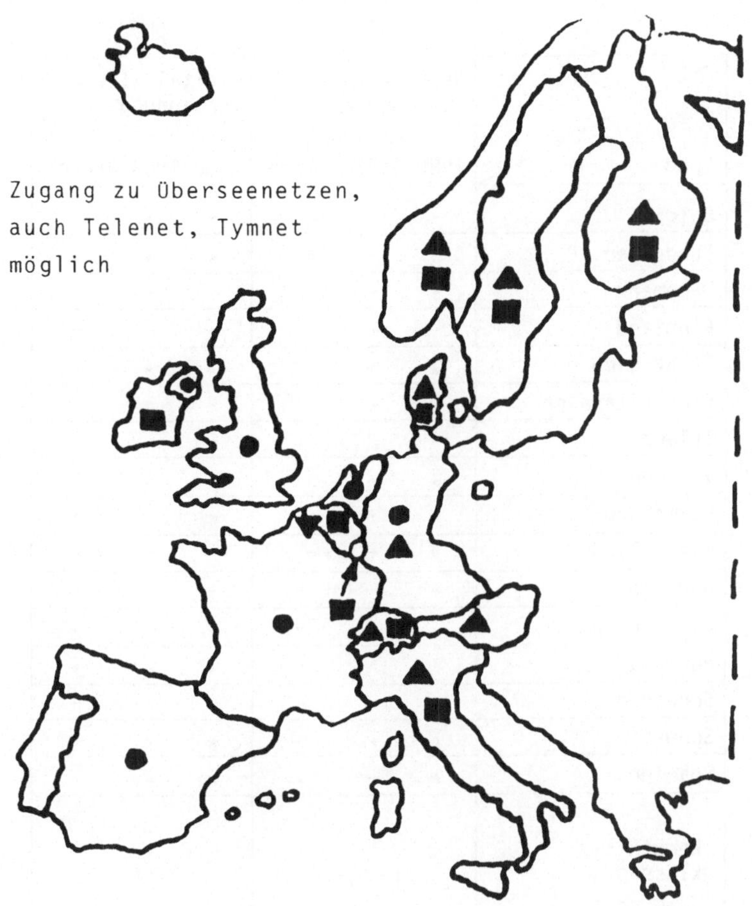

Bild 6

Netz / Land	MS — Nachrichten-speicherverm. 1980	1983	danach	DL — Digitale feste Verbindung 1980	1983	danach
Belgien	-	-	o	-	-	o
Bundesrep. D	-	-	o	*	*	*
Dänemark	-	-	-	-	-	-
Finnland	-	-	-	-	-	-
Frankreich	*	*	*	*	*	*
Großbritannien	*	*	*	-	o	o
Irland	-	-	-	-	*	*
Italien	*	*	*	*	*	*
Luxemburg	-	-	-	-	-	-
Niederlande	-	-	o	-	-	-
Norwegen	-	-	-	-	-	-
Österreich	-	-	-	*	*	*
Portugal	-	o	o	*	*	*
Schweden	-	-	-	-	-	-
Schweiz	o	*	*	*	*	*
Spanien	*	*	*	*	*	*
Benutzer-klassen nach CCIT	1 - 7			1 - 7		

* verfügbar

o Einführung erwogen, in Untersuchung

- nicht verfügbar, bisher nicht vorgesehen

PLÄNE FÜR DIE EINFÜHRUNG VON ÖFFENTLICHEN
DATENNETZEN IN CEPT - LÄNDERN

MS : message switching
DL : dedicated lines

Bild 7

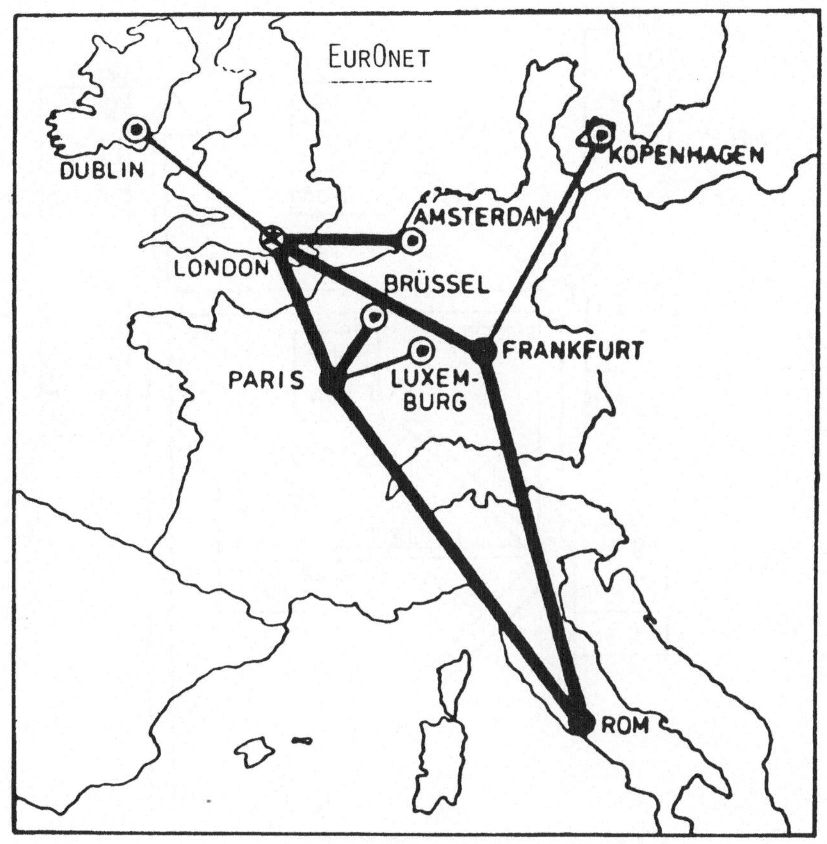

● DATENPAKETVERMITTLUNGSSTELLE

◉ KONZENTRATOR (ENTFERNTER ANSCHLUSSPUNKT)

★ NETZKONTROLLZENTRUM MIT PAKETVERMITTLUNGSSTELLE

▬ 48 KBIT/S − LEITUNG

▬ 9,6 KBIT/S − LEITUNG

EurOnet − Konfiguration

1980/81 EurOnet − Dienste auch in den Ländern
CH, E, S, N

Bild 8

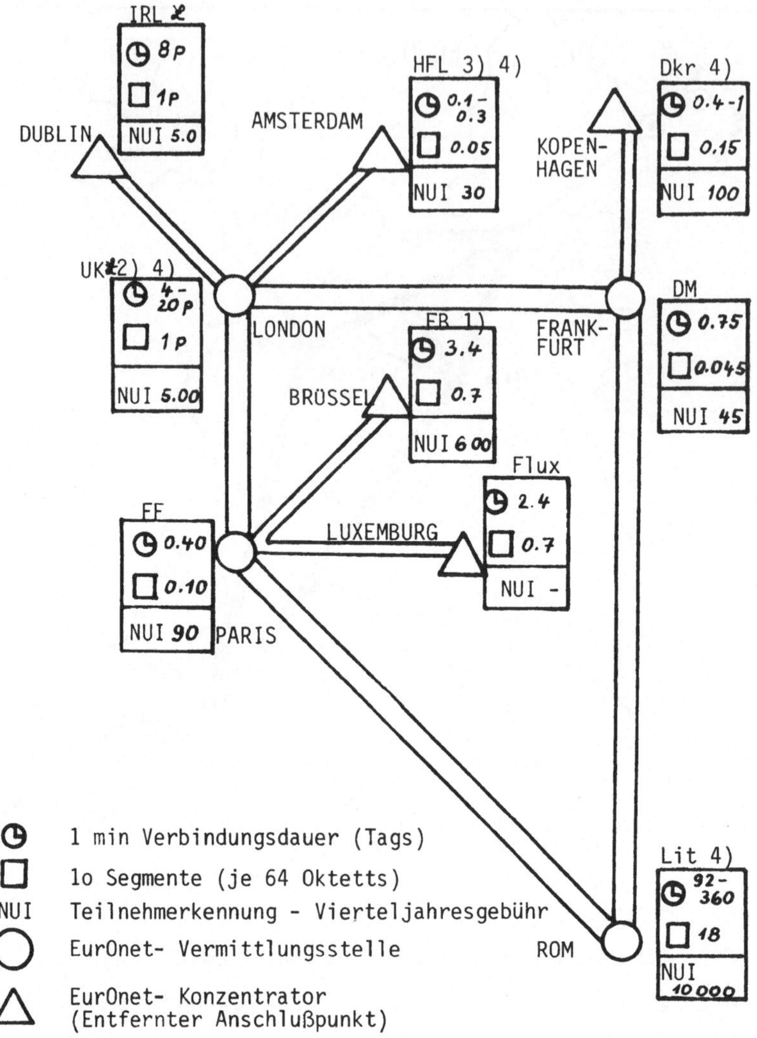

1 min Verbindungsdauer (Tags)

10 Segmente (je 64 Oktetts)

NUI Teilnehmerkennung - Vierteljahresgebühr

○ EurOnet- Vermittlungsstelle

△ EurOnet- Konzentrator
(Entfernter Anschlußpunkt)

Anmerkungen:

1) Besondere Zuschlagsgebühren
2) (einmalig bzw. periodisch)
3) Besondere Zuschlagsgebühr (je Verbindung)
4) Niedrigster - höchster Wert abhängig von der Entfernung

GRÖSSENORDNUNG EINIGER GEBÜHRENELEMENTE

Bild 9

gegeben ist. Die Umrechnung der einzelnen Gebührenelemente spiegelt nämlich ziemlich genau die europäischen Wechselkurse wider. Zum jetzigen Zeitpunkt werden die Teilnehmer direkt an Euronet angeschlossen (Bild 10). Mittelfristig werden die bereits erwähnten nationalen PS-Netze Zubringeraufgaben übernehmen. Gleichzeitig entstehen jedoch direkte internationale Bündel zwischen diesen PS-Netzen, sodaß langfristig (Bild 12) Euronet in das weltweite internationale Kommunikationsnetz integriert werden wird. Die notwendigen Voraussetzungen, nämlich CCITT-Definitionen für den Betrieb auf diesen Bündeln (z. B. X75), wurden bereits geschaffen.

1.3. Satelliten-Anwendung in Europa

In den vorhergehenden Vorträgen ist ausführlich auf die Satelliten-Technik in den USA eingegangen worden. Es soll daher auch der europäische Stand angesprochen werden, zumal in Europa einige beachtliche Satelliten-Anwendungen vorhanden sind. In Betrieb sind Forschungssatelliten der Projekte Symphonie und OTS (orbital test satellite). In einer weit fortgeschrittenen Projektierungsphase sind

TV-SAT (mehrere Staaten)

ECS (european communication satellite)

Télécom 1 (Frankreich)

Die wesentlichen Daten des Symphonie-Systems zeigt Bild 13. Die Verschiebung eines Satelliten von 11, 5° W auf 49° E im Jahr 1977 ermöglichte es Indien, für nationale Tests Zugriff auf einen Satelliten zu haben. Durch Konzipierung auch sehr kleiner mobiler Antennen war es möglich, in größerem Umfang mobile Einsätze zu erproben. Das Nutzungsprogramm umfaßt 4 wesentliche Bereiche:

technisch-wissenschaftliche Eyperimente
  Sat.-Funktionsprüfungen
  Rechnerverbund-Versuche
  Synchronisierung von Atomuhren

  Indien: nationale Versuche

Demonstrationen
  Salon du Bourget 75/77
  TELECOM Genf
  Internat. Industrie-Ausstellung Teheran 1976

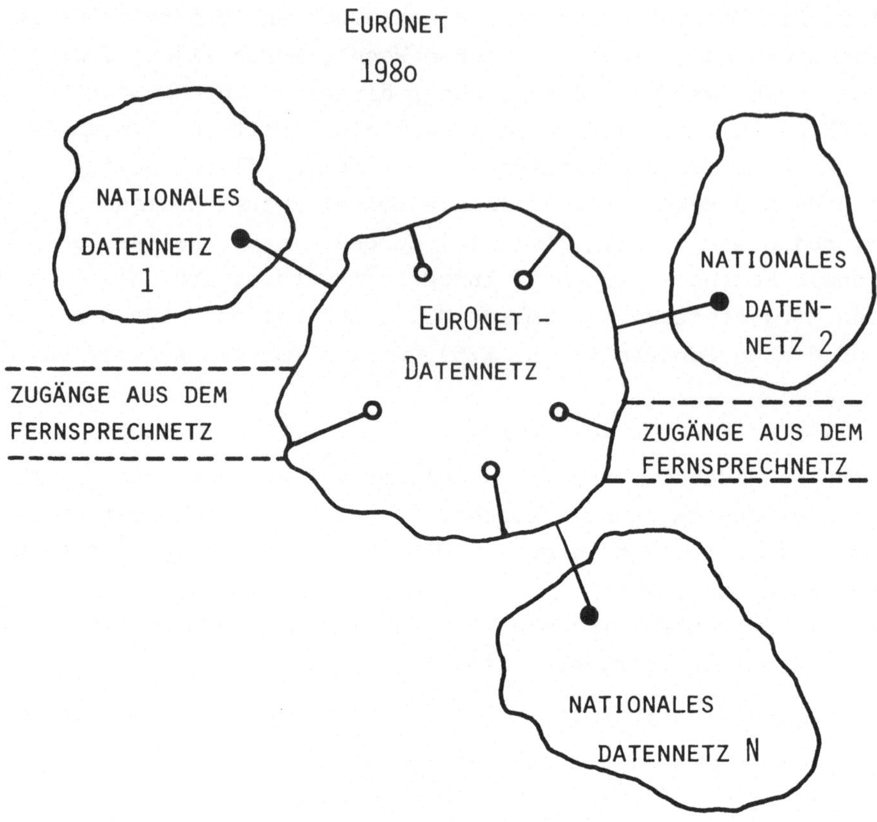

Bild 1o

Kulturelle und humanitäre Einsätze
   Hilfsaktionen des IRK
   Bildungsfernsehen in Afrika
   Einsätze für UNO/UNESCO

Betriebsversuche
Fernsprechen, Ton- und TV-Übertragung für
   Deutsche Welle (Kigali, Ruanda)
   ORTF → La Réunion
   ORTF → St. Pierre et Miquelon

Für **digitale** Versuche im Bereich der Satellitenübertragung wurde
5.78 der OTS (orbital test satellite) auf seine Position gebracht.
Die wesentlichen Daten zeigt Bild 14. Als Netzkontrollzentrum
(SCTS) wird die italienische Station Fucino eingesetzt. Weitere
große Erdefunkstellen haben Zugriff auf den breitbandigen A-Teil

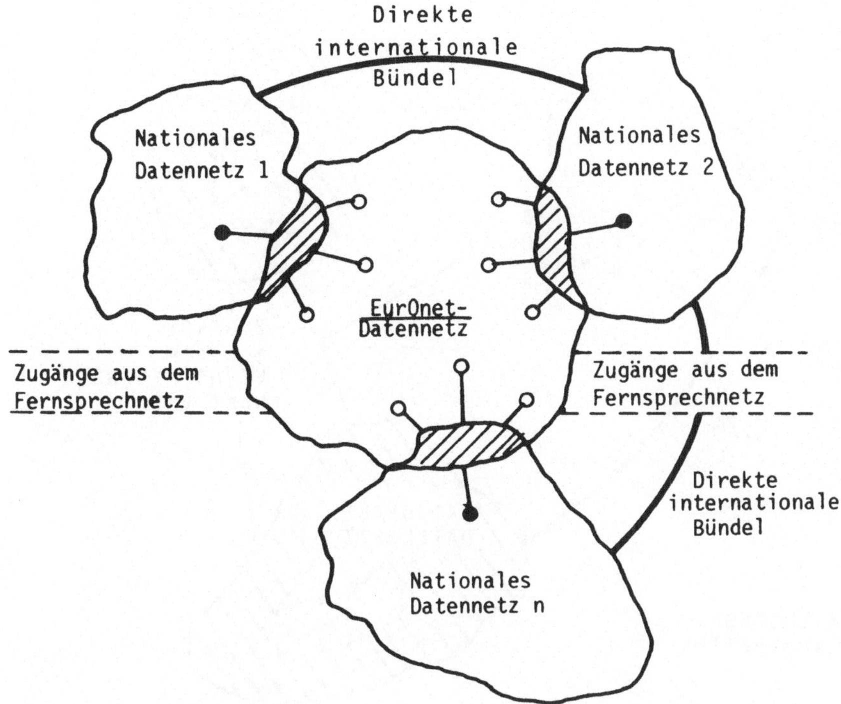

Bild 11

des Satelliten. Im Bereich der DBP wird hierfür die Erdefunkstelle
Usingen betrieben. Das Testprogramm besteht aus mehreren Teilen:

ESA-Versuche: Ausbreitungsmessungen, Verstärkerverhalten

Erprobung des Mehrfachzugriffs im Zeitvielfach
(Time Division Multiple Access, TDMA) mit 60 bzw. 120 MBit/s-
Signalen. Beteiligt sind ca. 10 Erdefunkstellen mit je 64
PCM 30-Rahmen, Takt 8 ms.

STELLA-Datenexperiment: Austausch wissenschaftlicher Daten
zu/von kleinen Erdefunkstellen am Standort der Beteiligten

SPINE-Experiment über kleine Erdefunkstellen:
  High Speed Facsimile-Übertragung
  Datenübertragung zwischen Rechnern
  Videokonferenz-Versuche

ZUKUNFT (LANGFRISTIG)

INTERNAT. PS - NETZ

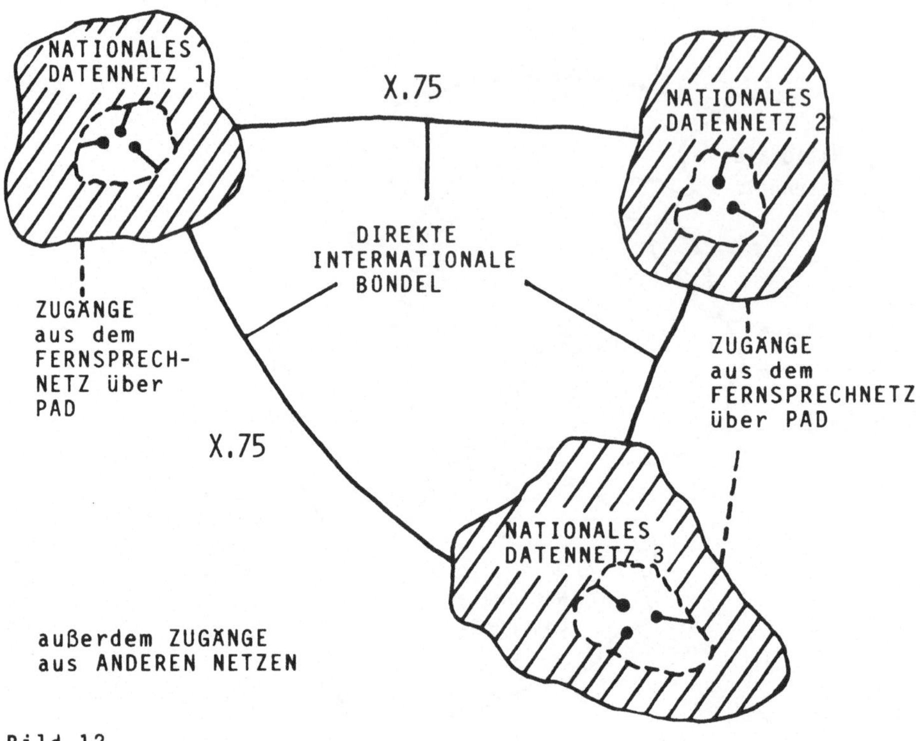

Bild 12

Von den PTTn werden noch TV-Übertragungsversuche (2 TV-Kanäle
über einen Verstärker) vorgenommen.

Die Fortschritte auf dem Satelliten-Sektor in Europa zeigt das
Bild 15. Es zeigt maßstabsgleich die Größenverhältnisse von OTS
und TV-SAT.

Die Planungen für den Einsatz von TV-Satelliten sind in Europa
sehr weit fortgeschritten. Entsprechende detaillierte Planungen
in den USA sind bisher nicht bekannt geworden. Hier besteht daher
für die europäische Industrie die Chance, eine führende Position
zu erlangen. Die grundsätzliche Funktion ist aus Bild 16 zu ersehen. Über nationale Richtfunkzubringer und Erdefunkstellen
werden die Satelliten mit vorerst 3 (später 5) Programmen ver-

SYMPHONIE (D/F)

Start 12.74/8.75 (11,5°W)
1977 Verschiebung eines Satelliten auf 49°E
Geplante Nutzung bis 1981
Frequenzber.: 4 GHz ↓ 6 GHz ↑

Antennen Ø m : 2,2 - 16
    fest und mobil

Feste Erdefunkstellen (16 m):
    2 TV-Farbkanäle oder 800 Telefonkanäle

Mobile Erdefunkstellen (3; 2,2):
    1 Telefon-, 1 Fernschreibkanal

Nutzungsprogramm:
Techn./wissensch. Experimente
Demonstrationen
Kulturelle und humanitäre Einsätze
Betriebsversuche

Bild 13

sorgt, die sie im 12 GHz-Bereich von den im Orbit zugewiesenen
Positionen zurückstrahlen. Die TV-Signale werden über Gemeinschafts-
antennenanlagen oder auch über kleine Parabolspiegel direkt den
Heimfernsehgeräten zugeführt. Diese rein nationale Versorgung durch
gebündelte Abstrahlung ist auf der internationalen Funkverwaltungs-
konferenz 1977 (WARC 77) festgelegt worden. Die wesentlichen Daten
des TV-SAT-Systems sind:

40 Kanäle im Bereich 11727,48 - 12475,50 $\overline{MHz}$,
Kanalabstand 19,18 $\overline{MHz}$

Orbitpositionen der Region 1 (Europa)
von 43^0 West bis 29^0 Ost, Abstand jeweils 6^0

Je Kanal und Position _zwei_ Polarisationsrichtungen

ORBITAL TEST SATELLITE
Start 5.78

OPT(orb. test progr.):ESA,interim Eutels.
Schwerpunkte: Digitale Anwendungen

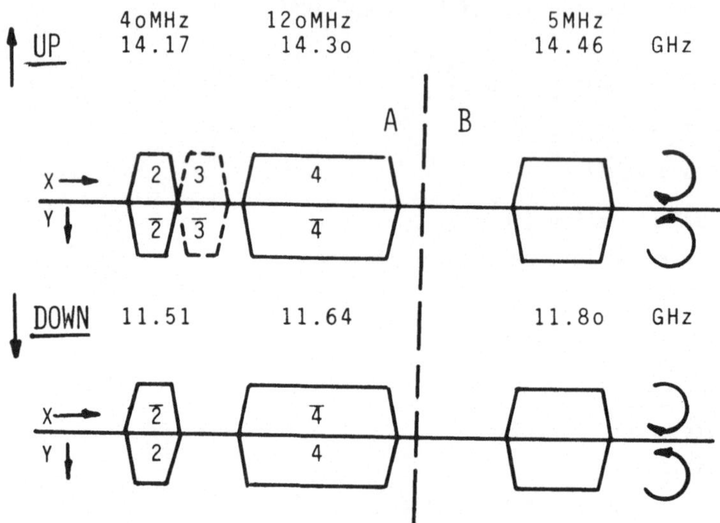

Kanäle 3/3̅ nur UP für Tests

FUCINO Sat. Contr. and Test Stat.

Weitere Erdefunkstellen: Goonhilly
                         Bercenay-en-Oth   für A   -Bereich
                         Usingen
                         ca.4o kleinere EFu-St für A+B-Bereich

Bild 14

Zuteilung von jeweils 5 Kanälen nach dem Schema
        1, 5, 9, 13, 17
        2, 6, 10, 14, 18
              bis
        24, 28, 32, 36, 40

Sendeleistung SAT 260 W/Kanal

Die von WARC 77 vorgeschriebenen Ausleuchtzonen zeigt Bild 17.
Hervorgehoben sind hier die Ausleuchtzonen (beams) von D, I und F.

Die deutsche Ausleuchtzone erfaßt noch Norditalien (Südtirol),
wo auch **heute offiziell das deutsche ZDF ausgestrahlt** wird. Auch
für Luxemburg ist ein eigener Satellit vorgesehen.

AEG-TELEFUNKEN

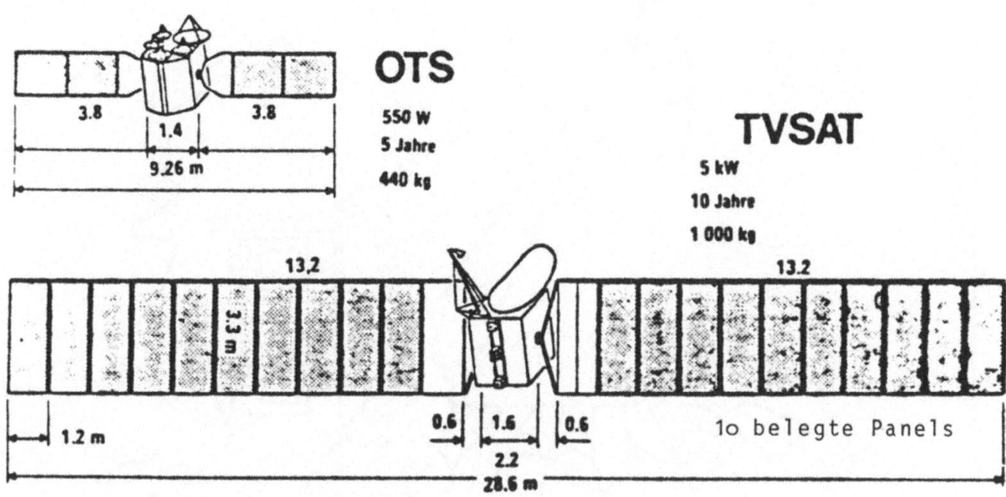

GRÖSSEN- UND LEISTUNGSVERGLEICH
REGIONALER NACHRICHTENSATELLIT U. HOCHLEISTUNGS -TVSAT

Bild 15

Die Vorteile der bereits erwähnten Kanalzuteilung zeigt Bild 18.
Die Länder A, D und CH werden Satelliten auf 19° West mit der
gleichen Position haben. Daher können in den relativ großen Über-
lappungsbereichen der einzelnen beams auch alle Programme ohne
Ändern der Position der Erd-Antenne empfangen werden. ^{D}as Gleiche
gilt für französisch sprechende Länder (F, L, B, NL).

Als fester Bestandteil eines digitalen europäischen Fernmelde-
systems soll 1983 der European Communication Satellite (ECS) auf
der Position 10° Ost in Betrieb genommen werden. Alle Betriebser-
fahrungen mit OTS werden hier einfließen. Die wesentlichen **Merkmale**
sind:

12 Transponder mit je 80 MHz Bandbreite

Frequenzbereiche down 11 - up 14 GHz

Als Zubringer/Abnehmer PCM/Systeme

TV-SAT

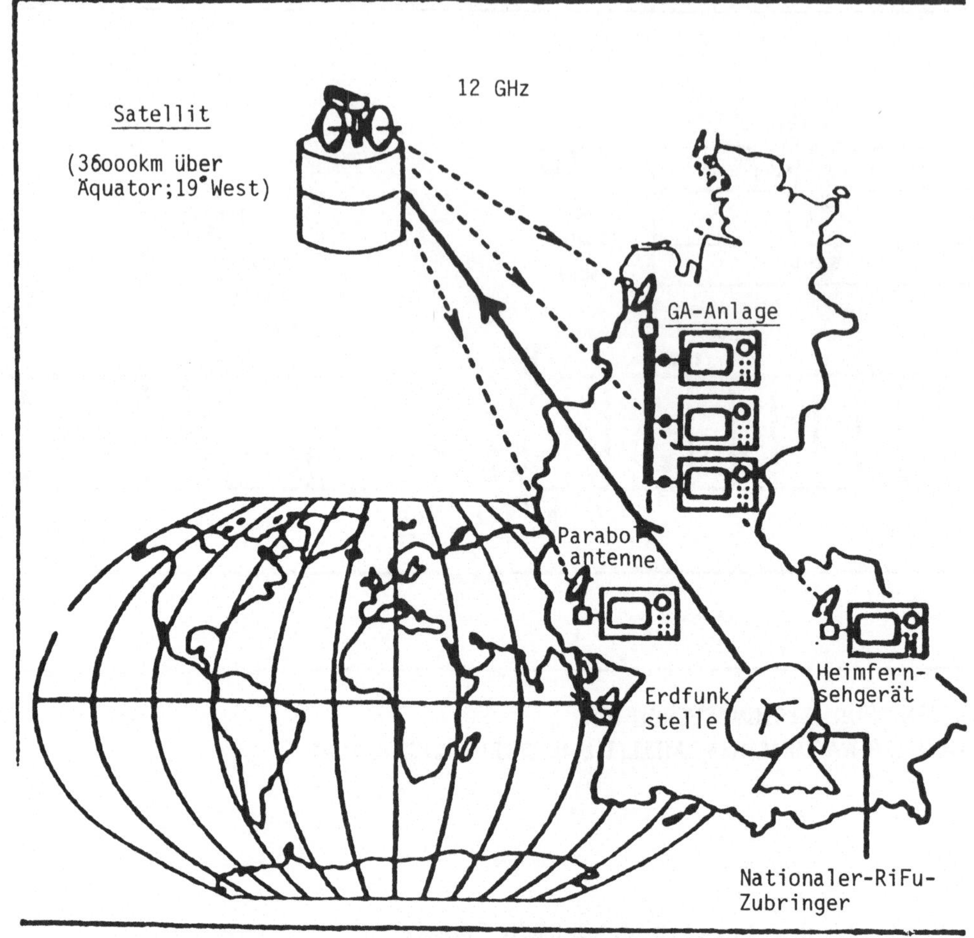

Andere Frequenzlagen: Intersat IV   4/6 GHz
                      Intersat   V 11/14GHz

Bild 16

In früheren Vorträgen war erwähnt worden, daß für Sprachübertragung
unter 2000 Meilen Bodenentfernung bzw. für Datenübertragung unter
500 Meilen keine Wirtschaftlichkeit zu erreichen ist. Diese für die
USA und Kanada durchaus üblichen Entfernungen werden in Europa nur
selten erreicht. Zur Erreichung eines wirtschaftlichen Satelliten-
betriebs werden daher zusätzliche Ausnutzungsverfahren angewendet:

TDMA wie bei OTS

DSI: Digital Speech Interpolation. Hierbei werden 240 terrest-
rischen Kanälen nur 120 Sat.-Kanäle zugeteilt. Bei jeder Sprach-

AEG – TELEFUNKEN

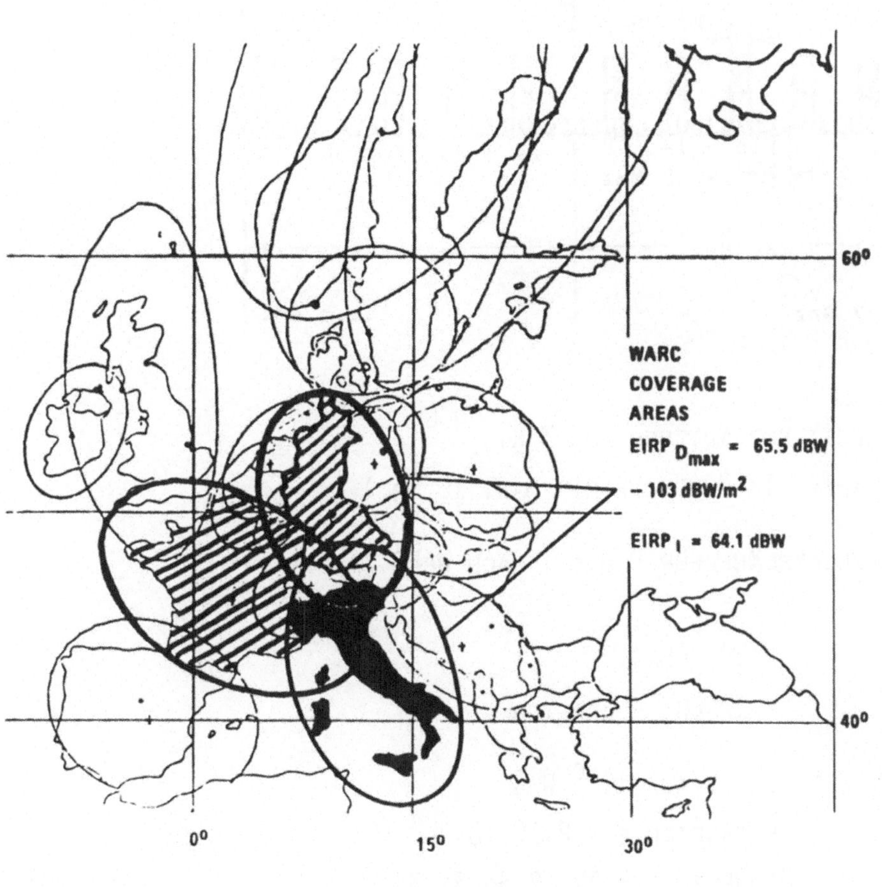

Ausleuchtzonen nach WARC 77

TV – SAT

Bild 17

pause wird der ^{S}at-Kanal sofort einer anderen Verbindung zugeteilt.
Bei weiterem Sprechen auf dem ersten Kanal wird dann im ms-Bereich
ein <u>anderer</u> gerade freier Sat.-Kanal zugeteilt. Die beteiligten
Erdefunkstellen müssen hierfür einen besonderen DSI-Modul enthalten. Ein ähnliches Verfahren (analog) ist auf Transatlantik-
kabeln schon betrieben worden (TASI, time assigment speech interpolation).

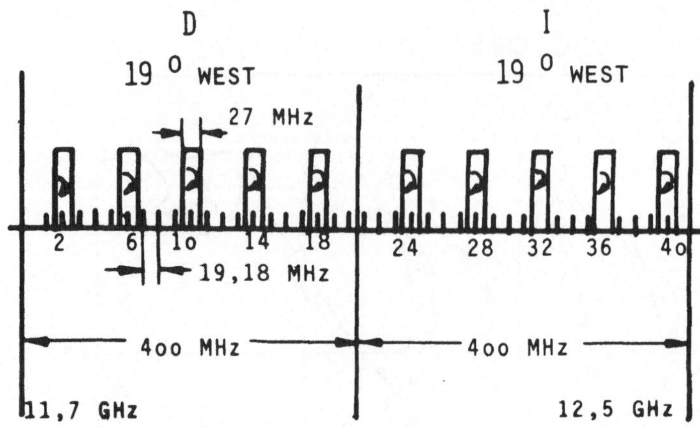

VORERST NUR NUTZUNG VON 3 K.

(2 x TV, 1x RUNDFUNK MIT MAX. 16 PR.)

FREQUENZPLÄNE FÜR D UND I NACH WARC 77

TV - SAT

12 GHz - BEREICH

WEITERE LÄNDER AUF 19 ⁰ W 2

ÖSTERREICH K 4,8,12,16,2o

SCHWEIZ   K 22,26,3o,34,38

LÄNDER AUF 19 ⁰ W 3

FRANKREICH, LUXEMBURG, BELGIEN

NIEDERLANDE

Bild 18

Multidestination (Vielzieltechnik). Hier werden die Kanäle
einer PCM 30-Gruppe unterschiedlichen Zielen zugeführt. Hierfür wurde das in Europa bereits angewendete Signalisierungs-
verfahren R 2 durch eine besondere digitale Version modifi-
ziert und in der Vermittlungstechnik des Satelliten zum Einsatz gebracht.

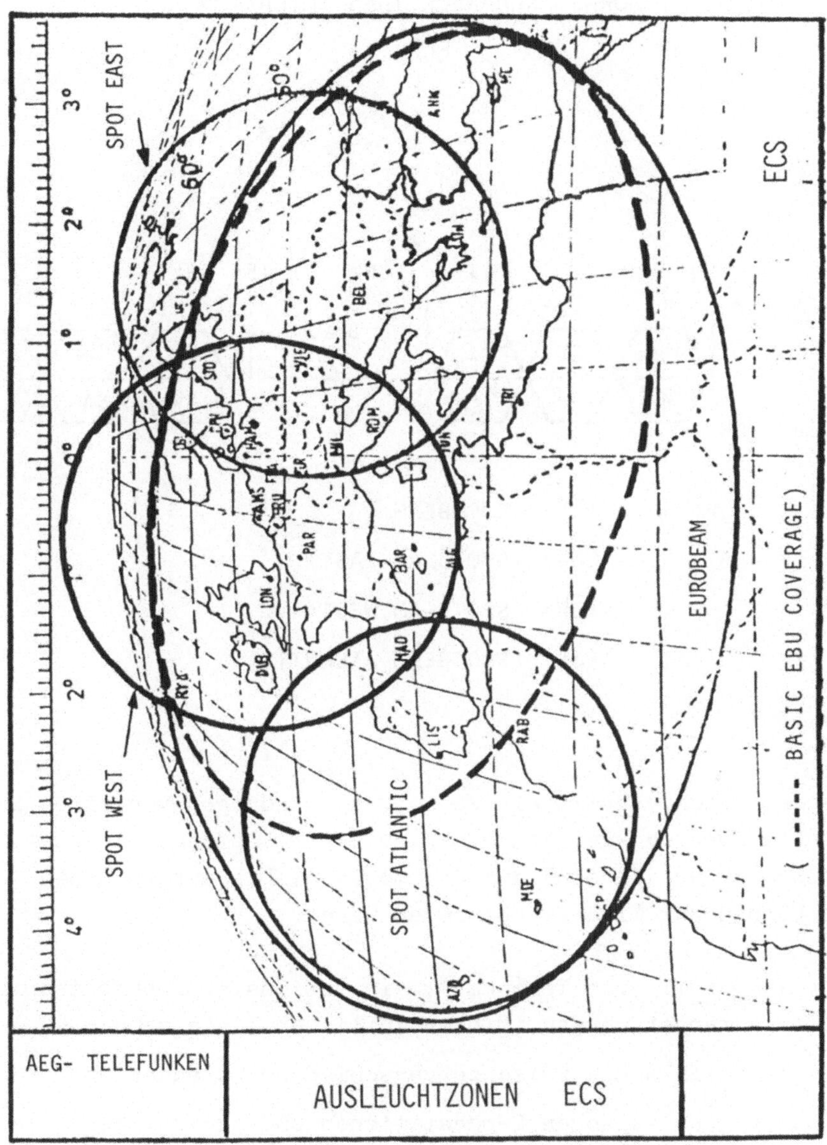

Bild 19

Die Ausleuchtzonen des ECS zeigt Bild 19. Neben dem Europa und
Nordafrika abdeckenden Eurobeam wird es für Zwecke der EBU
(European Broadcasting Union) eine etwas kleinere EBU-Keule geben.
Ferner werden drei kleinere Spotbeams (Atlantik, West, Ost) einen
Betrieb mit höherer Leistung oder kleineren Antennen ermöglichen.
Im Satelliten kann durch ein digitales Koppelnetz nun zwischen
bestimmten Transpondern, die vorrangig für diese einzelnen beams

DOWNLINKCHANNEL IDENTIFICATION
ECS

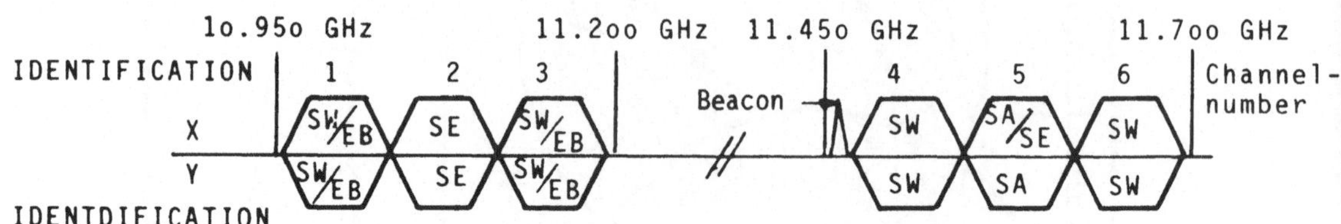

KEY: (EB) EUROBEAM
        (SE) SPOTBEAM EAST
        (SW) SPOTBEAM WEST
        (SA) SPOTBEAM ATLANTIC

Bild 2o

betrieben werden, **vermittelt** werden. Verbindungen von einer Sende-
antenne im Eurobeam-Bereich z. B. zu einer Empfangsantenne im
Atlantikbeam-Bereich sind dadurch möglich. Die vorgesehenen Zuord-
nungen können dem Bild 20 entnommen werden.

Im Jahr 1983 soll auch Télécom 1, ein nationaler französischer
Satellit gestartet werden. Die wesentlichen Ziele sind:

Schnelle Datenkanäle zwischen verschiedenen Standorten

TV-Übertragung zwischen Sendestationen und zahlreichen "kleinen"
Empfangsstationen

Fernsprech- und TV-Dienste zwischen F und "overseas"-départements

Hierfür werden Transponder ähnlich ECS eingesetzt. Um die oben
erwähnten Ziele zu erreichen, müssen 2 Frequenzbereiche benutzt
werden:

für Übersee down 4, up 6 GHz ($\hat{=}$ Symphonie)

für nationale Anwendungen (kleine Antennen in Ballungsgebieten)
down 12,6, up 14 GHz ($\hat{=}$ annähernd ECS)

Orbitposition 10° West.

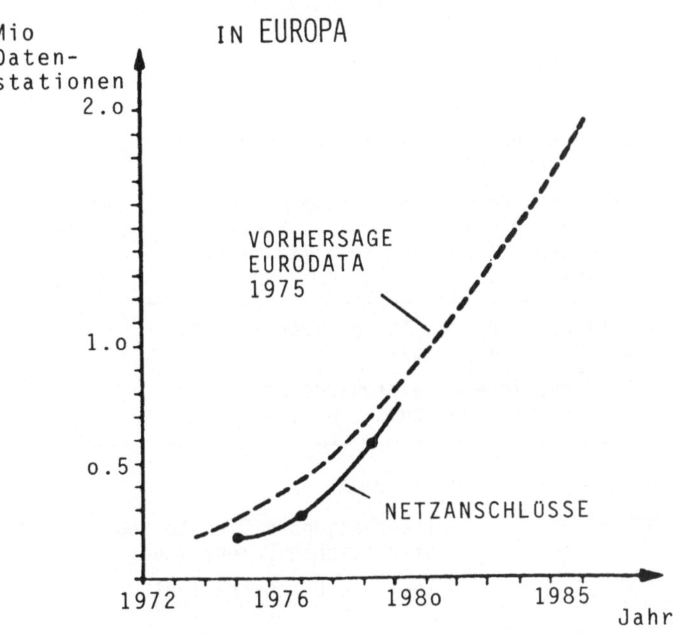

ENTWICKLUNG DER DATENSTATIONEN IN WESTEUROPA

Bild 21

Z. Z. laufen Verhandlungen zwischen F und anderen europäischen
Verwaltungen über die Anmietung bzw. Nutzung von Transpondern
dieses Satelliten. Es würden dann zwei unabhängige SAT-Systeme
mit annähernd gleichen Eigenschaften zur Verfügung stehen.

2. Datenübermittlung in Europa

Die Entwicklung der Zahl der Datenstationen in Europa zeigt das
Bild 21. Die Zahl der Netzanschlüsse ist gegenüber der ersten
Vorhersagen zwar etwas zurückgeblieben, zeigt aber dafür gegenüber der Vorhersage eine überproportional steigende Tendenz. Es
ist damit zu rechnen, daß 1985 mehr als 2 Millionen Netzanschlüsse
vorhanden sein werden.

Die Daten (und Text-) Übermittlung ist ohne Zweifel das wesentliche Element der sogenannten Value Added Networks (VAN). Für
VAN gilt folgende Definition:

ALLGEMEINE DIENSTE FÜR TLN

LEISTUNGSMERKMALE   CS

User facility	All user classes
1. Optional user facilities assigned for an agreed contractual period	Bestimmung, ob das Leistungsmerkmal international verfügbar sein muß oder kann
17 verschiedene Leistungsmerkmale wie z.B. Direktruf, geschl. Benutzergruppe, Anschluß-Identifizierung, usw.	
2. Optional user facilities requested by the DTE on a per call basis	
6 verschiedene Leistungsmerkmale wie Kurzruf, Rundsenden, usw.	

EMPFEHLUNG X.2 :   Internationale Dienste und Leistungsmerkmale in öffentlichen Datennetzen

Teil a) Durchschaltnetze

Bild 22

Alle Dienste/Merkmale, die über den für ein Netz vorgesehenen reinen Informationstransport hinausgehen, sind VALUE ADDED SERVICES

Hierzu zählen

allgemeine Dienste für Teilnehmer, definiert in CCITT X 2

Prozedur/Protokollanpassunegn

Umsetzung von in CCITT X.1 definierten Geschwindigkeiten (Benutzerklassen)

Speicherdienste

Netzunterstützung (z. B. mit vom Netzbetreiber bereitgestellten Rechnern)

Erst das Zusammenwirken dieser Funktionen (services) ergibt eine optimale Zusammenarbeit von INFO_Quellen und -Senken. Besondere Netze (VAN) sind hierfür jedoch nicht notwendig.

Die allgemeinen Merkmale für Durchschaltenetze zeigt Bild 22. Im 1. Teil sind 17 Merkmale festgelegt, die für ein Netz vorhanden sein müssen. Im Bereich der DBP z. B. sind im IDN nahe-

ALLGEMEINE TLN-DIENSTE
"DIENST"-MERKMALE   PS

Services	User classes of service		
	1 - 2	3 - 7	8 - 11
Virtual call service	E	FS	E
Permanent virtual circuit service	FS	FS	E
Datagram service	FS	FS	A

E = essential
A = additional
FS= further study

EMPFEHLUNG X.2 :   Internationale Dienste und Leistungsmerkmale in
                     öffentlichen Datennetzen

                     Teil b1) Dienste in Paketvermittlungsnetzen

Bild 23

zu alle Merkmale vorhanden. Für die einzelne Verbindung sind
z. Z. 6 verschiedene Merkmale definiert. Nach Definition und
Einführung dieser Merkmale wird es praktisch keine zusätzlichen
Kundenwünsche mehr geben.

Für die Dienste in paketvermittelten Netzen sind in Bild 23 die
Dienste definiert worden. Die Netze müssen bereits heute für
Verbindungen die Benutzerklassen 1,2 und 8 - 11 anbieten. Auch
für PS-Netze werden in Bild 24 wieder unter 1. Merkmale für das
eigentliche Netz sowie unter 2. für die Verbindung definiert.
Besonders zu erwähnen ist hier unter 1. die Durchsatzklasse, aus
der die volumenabhängige Tarifierung abgeleitet werden kann. Hin-
zu kommen jedoch noch unter 3. Empfehlungen für den Übergangsver-
kehr zwischen PS-Netzen und anderen Netzen. Diese PAD-Empfehlungen
(packet-assemby-disassembly) reichen bis in die Schnittstelle des
Endgerätesektors hinein. Auch diese Merkmale werden bereits heute
von mehreren europäischen Datennetzen angeboten (DATEX-P, Transpac,
EPSS). Für fest geschaltete Verbindungen (Bild 25) liegen die entsprechenden Empfehlungen ebenfalls vor.

Die Prozedur/Protokollanpassung bzw. die Verbindungen zwischen
unterschiedlichen Benutzerklassen nach CCITT X.1 können wieder
von verschiedenen Standpunkten aus betrachtet werden:

ALLGEMEINE TLN-DIENSTE

LEISTUNGSMERKMALE   PS

User facility	User classes of service					
	1-2			8-11		
	VC	PVC	DG	VC	PVC	DG
1. Optional user facilities for an agreed period	Bestimmung, ob das Leistungsmerk- mal international verfügbar sein kann oder muß, oder noch weiter studiert werden muß					
29 Leistungsmerkmale wie Durch- satzklasse, geschl. Benutzer- gruppen, usw.						
2. Optional user facilities on a per call basis						
13 Leistungsmerkmale wie Kurz- ruf, Rundsenden, usw.						
3. Optional user facilities provided by a PAD						
24 Leistungsmerkmale - hauptsächlich TTY-bezogen						

```
VC  = Virtual call
PVC = Permanent
        Virtual Circuit
DG  = Datagram
```

Empfehlung X.2:   International Dienste und Leistungsmerk-
                   male in öffentl. Datennetzen

                   Teil b2) Leistungsmerkmale in
                             Paketvermittlungsnetzen

Bild 24

Aus Sicht der Technik: Verbindungen zwischen Teilnehmern der
Durchschaltenetze (CS, asynchron) und Teilnehmern von z. B.
PS-Netzen, also PAD-Funktion nach CCITT X.3 (28,29).

Aus Sicht der Dienste: Verbindungen zwischen Teilnehmern
verschiedener Dienste, z. B.
Telex/Teletex, Bildschirmtext/Teletex u. a.

Diese Umsetzungen sind zwingend, weil z. B. Teletex betrieben
wird

im Bereich DBP im CS-Netz (IDN)

in F im PS-Netz (Transpac)

in einigen Ländern im Fernsprechnetz

Die bereits mehrfach erwähnten, von CCITT definierten Benutzerklassen sind aus Bild 26 zu ersehen. Die Empfehlung sieht 11

ALLGEMEINE TLN-DIENSTE

LEISTUNGSMERKMALE   DL
(fest geschaltete Verbindungen)

User facility	User classes of service	
	1-2	3-7
1. Point-to-point	E	E
2. Centralized multipoint	A	A

E = essential
A = additional (muß international nicht in jedem Fall ange-
    boten werden werden)

Empfehlung X.2:   International Dienste und Leistungsmerk-
                   male in öffentl. Datennetzen

                   Teil c) Internationale Mietleitungen

Bild 25

BENUTZERKLASSEN NACH CCITT

User class of service	Data signalling rate and code structure	Address selection and call progress signals
1	300 bit/s, 11 units/character	300 bit/s, I.A. No. 5 (11 units/character)
2	50-200 bit/s, 7,5-11 units/character	200 bit/s, I.A. No. 5 (11 units/character)

a) für Start-Stop-Betrieb

3	600 bit/s	600 bit/s,I.A.No.5
4	2400 bit/s	2400 bit/s,I.A.No.5
5	4800 bit/s	4800 bit/s,I.A.No.5
6	9600 bit/s	9600 bit/s,I.A.No.5
7	48000 bit/s	48000 bit/s,I.A.No.5

b) für synchronen Betrieb

8	2400 bit/s	2400 bit/s
9	4800 bit/s	4800 bit/s
10	9600 bit/s	9600 bit/s
11	48000 bit/s	48000 bit/s

See Recommendation X.25 for user packet format

c) für paketorientierten Betrieb

Empfehlung X.1:   Geschwindigkeitsklassen in
                   öffentlichen Datennetzen

Bild 26

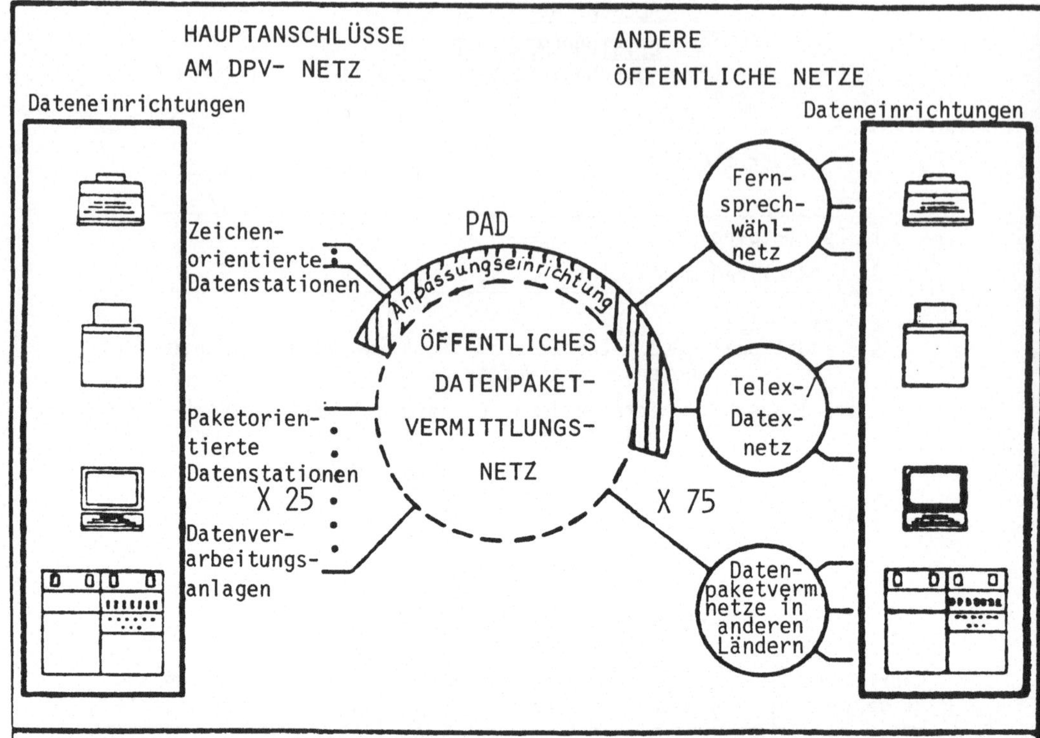

Bild 27

"user classes of service" vor:

1 - 2 für asynchrone Netze

3 - 7 für synchrone Netze (CS)

8 - 11 für paketorientierten Betrieb (PS).

Jede europäische Verwaltung wird diese Klassen so weit wie möglich realisieren bzw. bei Nichtanbieten bestimmter Klassen den Netzübergang sicherstellen.

Einen Überblick über die aus heutiger Sicht notwendigen Übergänge zeigt das Bild 27. Sämtliche Empfehlungen für die gezeigten Übergänge liegen vor.

Aus Anlaß der Inbetriebnahme des DATEX-P-Netzes (DBP) im Juli 1980 sollen die wesentlichen Merkmale vorgestellt werden:

    Datex-P 10: Basisdienst X 25
        2,4 - 48 KBit/s
        Leistungsmerkmale X.2

Datex-P 20: PAD-Funktionen, Übergang zu anderen Netzen
  Anschluß zeichenorientierter Datenendeinrichtungen
  Einwahl aus F-Netz und Datex-L

Datex-P 32: Anschluß nicht intelligenter Bildschirmstationen
und DVA, kompatibel mit IBM 3270

Datex-P 33: wie 32, jedoch kompatibel mit Siemens 8160

Datex-P 3.....   ?

Langfristige Zielsetzung ist es, durch Verarbeitungsfunktionen
im Netz (P 32, P 33) den Kunden der herstellerorientierten Netze
einen größeren Freiraum für ihre Planungen zu verschaffen. Es ist
diesen Kunden dann möglich, z. B. mit ihren heutigen Einrichtungen
auch "fremde" DVA oder umgekehrt zu erreichen.

Das Thema <u>Speicherdienste</u> soll nur kurz angesprochen werden. VAN
sollen vollständige Nachrichten transportieren. Es soll dabei
keine Strukturierung (z. B. Envelopes 6 + 2 oder 8 + 2) vorge-
schrieben werden. Typische Anwendungen wären:

Zwischenspeichern und späteres gezieltes Absenden, z. B.
elektronische Post im Sinne der ITT-Studie

Rundsenden nach Zwischenspeicherung

Nachrichtenumlenkung, Anrufumleitung

Diese Merkmale sind bereits in zahlreichen europäischen Anwendungen vorhanden (öffentliche Technik und Nebenstellenan-
lagen).

Für eine <u>Netzunterstützung</u> sind Protokollvereinbarungen für an-
wendungspezifische Funktionsschichten Voraussetzung, z. B. durch
einheitliche höhere Kommunikations-Protokolle (EHKP) oder die
Definition der 4 Ebenen <u>oberhalb</u> CCITT X.25. Zweck ist die Ver-
lagerung bestimmter, von Endeinrichtungen <u>häufig</u> benötigter
Leistungsmerkmale in das Netz, wo dann nur <u>einmaliger</u> Pro-
grammieraufwand für zahlreiche Anwendungen nötig wird. Beispiele
hierfür sind:

Datex-P 3 ... bei DBP (und anderen europäischen Verwaltungen)

Bildschirmtext mit Anschluß von unterschiedlichen Info-Lieferanten

Es entsteht nun die Frage: Sind

VALUE ADDED NETWORKS ein Schlagwort ?

Value added services als Erweiterung des Angebots in CCITT-standardisierten Netzen sind Zielsetzungen der CEPT-Verwaltungen in Europa. Dieses Ziel kann durch bilaterale Vereinbarungen Zug um Zug erreicht werden. Einen Terminplan hierfür enthält die Studie der CEPT "Öffentliche Datennetze" im Teil "internationales Zusammenwirken (©1979, CEPT/Eurodata Foundation).

Daher sind getrennte Sondernetze in Form der in den USA vorhandenen VAN nicht notwendig. Der eigenständige Begriff VAN ist tatsächlich ein Schlagwort.

3. Dienste in Europa

Als Schwerpunkt sollen hier die Varianten der Textkommunikation behandelt werden. Auch hier stoßen wir wieder auf das Schlagwort "electronic mail". Die Definition sagt aus:

Übermittlung von Texten/Vorlagen aller Art auf elektrischem Wege, wobei Sender/Empfänger sein kann:

Büromaschine
Bildschirm
Faksimilegerät

Für einige Dienste gab es in Europa Entwicklungsschwerpunkte:

Videotex interactive in GB
Teletex in D
Videotex broadcast GB/F
Telefax D (und andere Staaten)

Leider ist weltweit eine Dienst-Bezeichnungs-Euphorie ausgebrochen. In Bild 28 befinden sich vor den einzelnen Namen Kennungen, wobei die Bezeichnungen mit derselben Kennung auch das Gleiche bedeuten. Die in Schrägschrift angedeutete "electronic mail" hat nun wiederum mit dem Ganzen etwas zu tun. Es wird höchste Zeit, daß CCITT für die verschiedenen Dienste international vereinbarte Empfehlungen herausgibt. Hierbei könnte der jeweilige nationale Name durch Nachsetzen sogar erhalten bleiben:

videotex interactive/Bildschirmtext für D
videotex interactive/PRESTEL für GB u. a.

Als erster Dienst soll Videotex broadcast, ein Dienst in Funknetzen (Breitband) behandelt werden. Die Situation in Europa ist wie folgt:

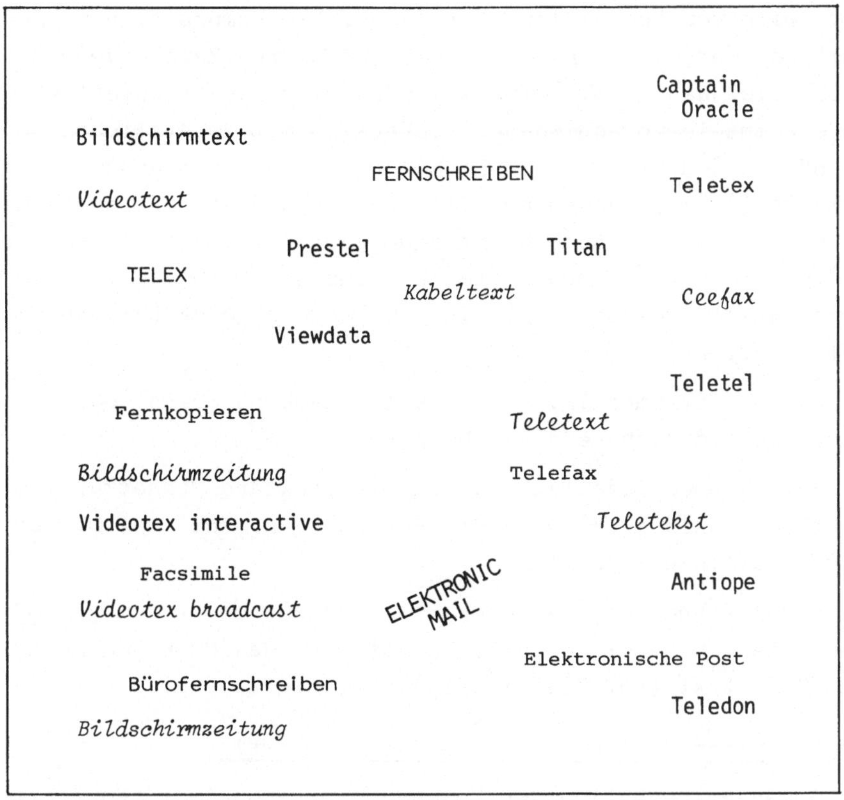

Bild 28

GB, öffentlicher Betrieb nach umfangreichen Versuchen mit CEEFAX/
ORACLE (BBC/IBA). Jetziger Name UK-Teletext mit Ausnutzung von
2 Zeilen des Videosignals. Hierdurch können ca. 100 Seiten angeboten werden. Zur Anwendung kommt die zeilengebundene Übertragung und die Direktmethode. Hiermit sind folgende Nachteile
verbunden: auch halb leere Zeilen müssen bei der Datenübertragung
die entsprechenden Leer-Zeichen enthalten, jeder Bitkombination ist
ein Zeichen direkt zugeordnet. Für den Zeichengenerator besteht
zwar eine eindeutige Zuordnung zwischen Bitkombination und abzu-
bildendem Zeichen, es muß aber für die einzelnen Alphabete insgesamt ein sehr großer Zeichenvorrat bereit gehalten werden.

D, Versuchsbetrieb über die TV-Sender des 1. und 2. Programms
seit 01.06.80. Es werden durch Ausnutzung von 2 Zeilen ca. 75
Info-Seiten angeboten. Info-Lieferanten (Betreiber) sind
ARD, ZDF, Zeitungsverleger, Wetterdienst u. a.

F, Versuchsbetrieb mit nicht zeilengebundener Übertragung und Kompositionsmethode. Durch direkte Adressierung werden keine ganzen

oder Teile von Leerzeilen übertragen. Die Kompositionsmethode
kommt bei der Codierung mit einem geringeren Zeichenvorrat aus.
Am Beispiel des e soll dieses Verfahren näher erläutert werden:
das e kommt in verschiedenen Alphabeten mit zusätzlichen "Sym-
bolen" (é, è u.a.) vor. Ähnliches gilt für viele andere Zeichen-
elemente (å, ø). Werden nun diese Symbole durch getrennte Zeichen
definiert, so kann nach Übertragung der Bitkombinationen für e
und für ' der Zeichengenerator hieraus das richtige Zeichen zusammensetzen (komponieren). In F werden folgende Leistungsmermale
realisiert:

D.R.C.S. (dynamically redefinable character sets) oder auch
down loaded alphabets (siehe oben)

"pay-T.V über Videotex": durch einsteckbare Magnetkarte mit
aktiver Logik (ROM) wird die Berechtigung zum Empfang bestimmter
Info-Seiten überprüft.

Ein vollständiger TV-Videokanal wird zur Übertragung großer
Info-Mengen bereitgestellt. Damit wird das Funk-Gegenstück
zu Kabeltext (auf Koax-Kabeln) realisiert.

Nun zu Videotex interactive (Bildschirmtext).

Die Aussagen bezüglich der Zeichengenerierung aus dem letzten
Abschnitt gelten auch für diesen Dienst, weil der TV-Empfänger
für beide Dienste als Datenendeinrichtung verwendet wird. Für den
"Bildaufbau" sind 3 Verfahren bekannt:

In Europa wird das Alpha-Mosaik-Verfahren verwendet. Hierunter
wird bei "Bild"-Darstellungen das punktweise Zusammensetzen der
Bildschirminformation verstanden, wobei die "Punktgröße" in Form
eines Quadrates relativ groß ist. Die Empfehlungen lassen die
Direkt- und die Kombinationsmethode zu, wobei für bestimmte
Zeichengruppen Kompatibilität gegeben ist. Die Vorschrift ISO 646
kennt ca. 300 α-numerische Zeichen, die bei der Direktmethode
16 Zeichensätze füllen würden. Die Kompositionsmethode kommt mit
2 Zeichensätzen aus: 1 Basissatz mit freien Plätzen für nationale
Anwendungen; 1 Satz mit allen diakritischen und Sonderzeichen. Mit
beiden Sätzen ist die Darstellung von 512 unterschiedlichen Zeichen
möglich.

In Übersee werden folgende Verfahren verwendet:

α - geometric (z.B. Kanada). Hier werden die einzelnen Zeichen
(ausgenommen Text) aus geometrischen Elementen (Kreiselemente,

Geraden mit unterschiedlichen Winkelstellungen u.a.) zusammenge-
setzt. Da auch der Durchmesser bei Kreiselementen wählbar ist,
können geometrische Figuren besser abgebildet werden. Der Zeichengenerator (einschl. Mikroprozessor) hierfür ist aufwendiger.

α- photographic (z.B. Japan). Hier erfolgt eine echte Ab-
tastung der Vorlagen, die dann in definierte Bitkombinationen um-
gesetzt werden, die in begrenzter normierter Zahl vorhanden sind.
Der Zeichengenerator setzt hieraus wieder die entsprechenden Zeichen zusammen. Besonders für die japanische Bilderschrift ist
dieses Verfahren natürlich optimal geeignet. Der Zeichengenerator
ist allerdings wieder aufwendiger, andererseits gestattet es
dieses Verfahren, den japanischen Videotex_Teilnehmern die Nachrichten in der gewohnten Weise zu übermitteln.

Die grundsätzliche Konfiguration des Netzes zeigt Bild 29. Über
das öffentliche Fernsprechnetz wird mit einer standardisierten
Datenübertragung (in D z.B. 1200/75 bit/s) die Bildschirmtextzentrale
erreicht. Nach Umschaltung auf Datenbetrieb können nun über Fern-
bedienung des TV-Empfängers mit 75 bit/s Befehle an die Zentrale
übermittelt werden. Die Rückrichtung mit Übermittlung der für den
Zeichengenerator vorgesehenen Bitkombinationen wird mit 1200 bit/s
betrieben.

GB: öffentlicher Betrieb mit der Direktmethode, zeilengebunden.
Die Bildschirmtextzentralen (BTZ) können nur aus der eigenen
Region ("Ortsverkehr") erreicht werden. Z. Z. sind in Betrieb
4 (10) nicht gedoppelte BTZ in London (in Klammern Stand
Ende 1980)

3 (6) gedoppelte BTZ in Birmingham, Edinburgh, Manchester
(Leeds, Cardiff, Chelmsford)

Zusammen ca. 4000 Teilnehmer mit ca. 1000 BTZ-Eingängen.

Ferner werden bei geringem Verkehrsaufkommen 48 KBit/s-
Multiplexer als abgesetzte Zugangspunkte verwendet, z. B.
Glasgow (Edinburgh).

Zentrales update-Zentrum in London

Jede BTZ enthält ca. 250.000 Seiten

Erreicht werden Ende 1980 die Regionen, in denen 65 %
der möglichen Teilnehmer vorhanden sind.

Durch das zentrale updaten ist der Änderungsdienst relativ auf-
wendig, weil alle Änderungen über ein besonderes Netz in alle

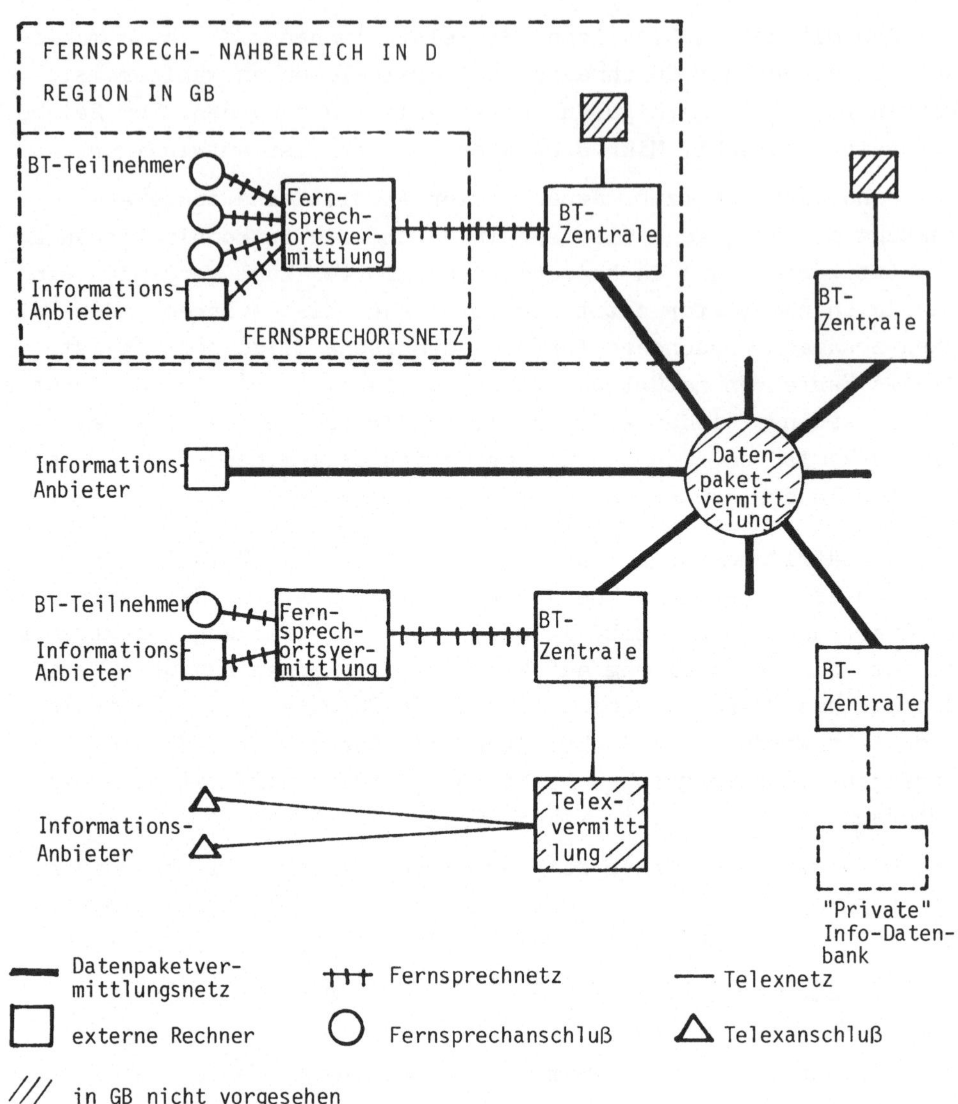

GESAMTSYSTEM BILDSCHIRMTEXT

Bild 29

BTZ, die immer <u>alle</u> den vollständigen Info-Vorrat enthalten müssen, eingegeben werden müssen.

D, Versuchsbetrieb seit 01.06.80 mit den **BTZ** Neuß (Bereich Düsseldorf) und Berlin. Vorgesehen ist

generelle Erreichbarkeit zu Nahdienstgebühren (8/12 Minutentakt)

Rechnerverbund zwischen BTZ über das PS-Netz (Datex-P) der DBP. Dadurch können selten benötigte Informationen oder Informationen, die durch Dritte ("Private", Bild 29) bereitgestellt werden, über dieses PS-Netz in die vom Teilnehmer angesteuerte BTZ in sehr kurzer Zeit bei Bedarf übertragen werden. Bei diesem Verfahren werden nicht alle Daten in allen BTZ abgespeichert.

F, Versuchsbetrieb TELETEL "Velizy".

Netzstruktur wie in D, Zubringer F-Netz, Rechnerverbund über Transpac

In F wird zeilenunabhängiges Verfahren mit 16 bit-Struktur verwendet.

<u>Picture Prestel</u> wurde durch BPO im März 1980 angekündigt und auf dem CCITT-Symposium Montreal 1980 (5.80) über interkontinentale Leitungen demonstriert. Bei diesem Verfahren kann 1/9 der Bildschirmfläche an <u>beliebiger</u> Stelle mit punktweise aufgebautem Farbbild in Video-Qualität belegt werden. Hierfür ist im Empfänger neben dem Zeichengenerator ein zusätzlicher Mikroprozessor mit 24 K-Speicher notwendig. Die Vorlage wird abgetastet, über DPCM codiert und in der BTZ abgespeichert. Die Anforderung des Bildes erfolgt durch zusätzlichen Befehl "press #".

Die Bildaufbauzeiten sind bei

1, 2 KBit/s   ca. 60 Sekunden
4,8   "   ca. 15   "
64,0   "   ca. 1 Sekunde.

Eine optimale Bildaufbauzeit wird erst in Digitalnetzen mit dem normierten 64 KBit/s-Kanal möglich sein.

Folgende Anwendungen werden erwartet:

Fernbestellung (mail order), Darstellung von Katalogangeboten in Text <u>und</u> Bild

Sicherheitsanforderungen, z. B. Kreditkartenprüfung durch Ab-
rufen der Unterschriftsprobe und Bild

Fernunterricht

Die Einführung ist z Z. noch nicht möglich wegen

relativ hoher Speicherkosten in BTZ und TV-Empfänger

Fehlens von CCITT-Empfehlungen

Einsatz erfolgt daher wahrscheinlich zuerst bei geschlossenen Be-
nutzergruppen. Der Vorteil ist, daß dem bereits vorhandenen Grund-
system PRESTEL (α- Mosaik) der Teilbereich Picture Prestel (α-
photographic) überlagert werden kann, weil die Übertragungsprozeduren
gleich sind.

Die Einführung des neuen Dienstes Teletex (Bürofernschreiben)
wurde maßgeblich durch Studien und Versuche in D beschleunigt.
Die wesentlichen Merkmale sind:

Neues Terminal mit vollständigem Zeichenvorrat einer Büro-
schreibmaschine

Unabhängige Lokalfunktion und

Unabhängige Übermittlungsfunktion zwischen Speichern der beteiligten Maschinen. Wegen dieser Merkmale muß eine eindeutige
Schnittstelle zwischen Lokal- und Übermittlungsteil vorhanden
sein (Bild 30).

Übergang Teletex/Telex muß von Anfang an möglich sein. Dadurch
können die zuerst nur in geringer Anzahl vorhandenen Teletex-
Kunden alle Teilnehmer des weltweiten Telex-Netzes mit den für
diesen Dienst vorgesehenen Merkmalen erreichen. Die Verwaltung
des Ursprungslandes der Teletex-Verbindung wird jeweils diesen
Übergang bereitstellen (value added service)

Weitere Übergänge vorstellbar, z. B. Teletex/Bildschirmtext.
Spezifikationen hierfür liegen noch nicht vor.

Einsatz in NStAnl. Daher muß auch der Übergang auf Fernsprech-
leitungen möglich sein.

Wenn ein flächendeckender Einsatz von Teletex erreicht werden soll,
wird dieser letzte Übergang eine große Bedeutung erlangen. Man
nähert sich damit dem Ziel der Kommunikations-Nebenstellenanlage,
einer Anlage, in der gleichberechtigt Sprache und Daten nebenein-
ander oder auch im Zeitmultiplex übermittel werden. Den grund-

sätzlichen Aufbau des Terminals zeigt Bild 30. Einige Hersteller
haben neben Tastatur und Drucker als Option auch schon einen Bildschirm vorgesehen. Die beiden Speicherbereiche (lokal, Übermittlungsteil) sind unabhängig voneinander. Der Lokalspeicher wird
auch für Zwecke der Schriftgutkorrektur verwendet. Einfaches
"Übertippen" im Entwurf führt zur "Reinschriftkorrektur". Außerdem können häufig wiederkehrende Redewendungen dem Speicher entnommen werden.

Die im Bereich der DBP vorgesehene Netzkonfiguration zeigt Bild 31.
Teletex wird mit 2400 bit/s im IDN abgewickelt. Fernsprech-NStAnl
werden über einen Umsetzer (TUFI) an das IDN angeschlossen. Der
Netzübergang zwischen IDN mit 2,4 KBit/s und Telexnetz mit 50 Baud
wird durch einen Umsetzer TTU geschaffen.

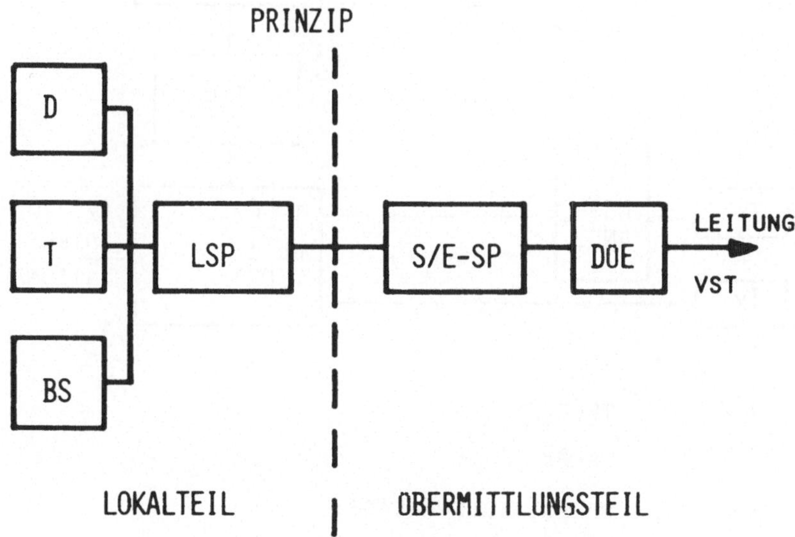

Bild 30

NETZKONFIGURATION FÜR
TELETEX/TELEX(DBP)

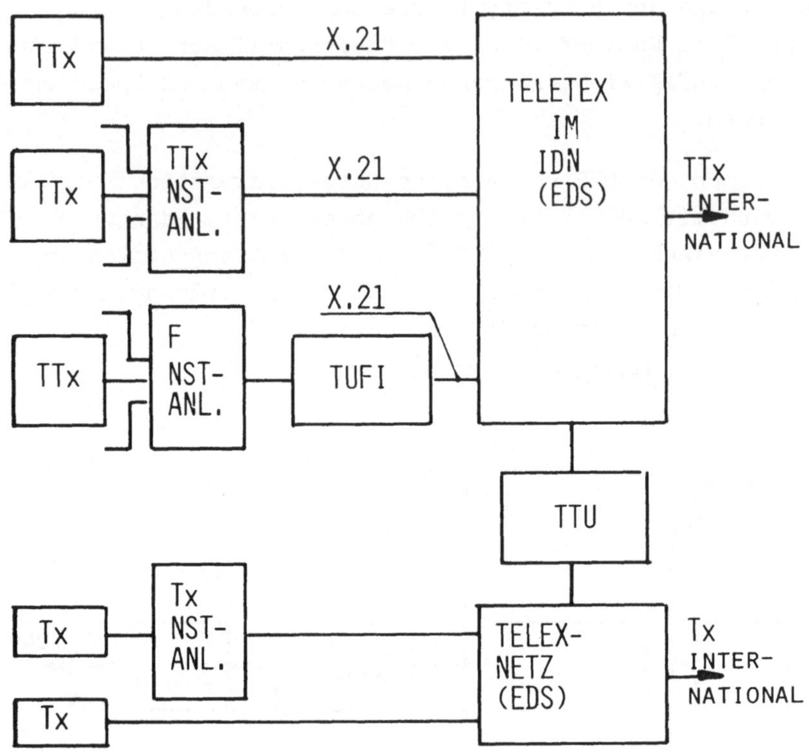

TTx: TELETEX

Tx: TELEX

NST-ANL.: NEBENSTELLENANLAGE

TUFI: TELETEX-UMSETZER FÜR FERNSPRECHEN/IDN

TTU: TELETEX/TELEX-UMSETZER

EDS: ELEKTRON. DATENVERM. SYSTEM

Bild 31

Die Einführung von Teletex ist wie folgt terminiert:

1976 - 80 umfangreiche Vorversuche in D mit erster Präsentation
auf der Hannover-Messe '80

6.80 endgültige CCITT-Empfehlung (Montreal, Stud.kom. I und VIII).

Ab 1.81 Versuchsbetrieb im öffentlichen Netz

Sommer 81 offizielle Aufnahme des Dienstes in D und S

Weitere Länder (CDN, F) folgen kurz danach

Teletex wird die Büroorganisation entscheidend verändern. Der in
vorherigen Vorträgen erwähnte Datenschutz, z. B. Schutz vor unbe-
rechtigtem Zugriff durch Dritte, ist auch im Dienst Teletex durch
die "password identification" bereits berücksichtigt worden. Ge-
schützte Nachrichten können an der empfangenden Maschine erst aus-
gelesen werden, wenn von dem Berechtigten das entsprechende password eingegeben wurde.

Das Fernkopieren ist in Europa schon seit langem mit zahlreichen,
untereinander nicht kompatiblen Geräten möglich. Wesentlicher
Inhalt des Telefax-Dienstes ist die Standardisierung kompatibler
Geräte der Gruppe 2 (3 min-Geräte). Mehrere Verwaltungen haben
diesen Dienst 1979 eingeführt. Der Stand ist wie folgt:

D: ca. 3000 Teilnehmer

F: ca. 1000   "

GB: kein CCITT-Standard

S: <1000

SF, DK, N: demnächst

Der Betrieb erfolgt im Fernsprechnetz. Zu diesem Dienst gehört auch
die Bereitstellung eines Teilnehmerverzeichnisses. Leider haben
einige Verwaltungen Teilnehmerverzeichnisse herausgegeben, in denen
auch Anschlüsse mit nicht-kompatiblen Geräten aufgeführt werden.
Damit wird das eigentliche Ziel dieses Dienstes natürlich nicht
erreicht. Als technisch wesentliches Merkmal soll nur die ·Auflösung
erwähnt werden. Sie beträgt 3,85 Linien/mm.

Bei zukünftigen Geräten der Gruppe 3 (1 min-Geräte) wird die
Auflösung 7,7/3,85 Linien/mm betragen. Auch hier ist wieder
der Betrieb im Fernsprechnetz vorgesehen. Im CCITT muß darauf
hingewirkt werden, daß eine Zwangskompatibilität zu den Geräten

der Gruppe 2 empfohlen wird, weil sonst die Einführung der
Gruppe 3-Geräte bei Vorhandensein einer großen Anzahl von Gruppe
2-Geräten kaum möglich sein wird.

In F wird z. Z. ein billiges "Volksfax, Gruppe 3" entwickelt.
Hierbei wird die Gruppe 2-Kompatibilität nicht vorgesehen.

Geräte der Gruppe 4 werden z. Z. für den Betrieb in Digitalnetzen spezifiziert. Hierbei soll das Protokoll dem Teletex-
Verfahren entsprechen, um Übergang Teletex/Telefax zu ermöglichen. Falls auch für Telefax die Übertragungsgeschwindigkeit
von 2,4 KBit/s gewählt wird, würde die Übertragung einer DIN A 4-
Seite 2 Minuten dauern. Ferner wird gesicherte Übertragung an-
gestrebt mit einer Auflösung ≥7,7 Linien/mm.

Es wird angenommen, daß 1985 in Europa bereits mehr als 100.000
Telefax-Teilnehmer (Gruppen 2 und 3) vorhanden sein werden. Für
die Zukunft wäre mit Geräten der Gruppe 4 noch die in Bild 32
angedeutete Formularaufteilung von Interesse. Bei diesem Verfahren könnten einerseits die firmenspezifischen Kennzeichen des
Formulars übertragen werden, andererseits aber für den Textteil
alle Vorteile der Schriftguterstellung mit Teletex genutzt werden. Im Dienst Teletex ist zur Realisierung dieser Kundenwünsche
vorgesehen, eine Formularkennung vorab zu senden, damit auf der
Empfangsseite das gewünschte "Formblatt" eingelegt werden kann.
Damit ist jedoch die Realisierung des in Bild 32 genannten Ver-
fahrens nur sehr unvollkommen möglich.

Einige wesentliche Komponenten des Begriffs "Electronic Mail"
wurden damit behandelt, ohne dabei den Anspruch auf Vollständigkeit zu stellen. Hieraus können nun geschlossene Dienste wie
z. B. die elektronische Post aufgebaut werden. Die Informations-
übertragung würde hierbei nach Telefax-Prinzipien erfolgen, die
Netzunterstützung mit Gebührenerfassung (Magnetkarten-Abbuchung ?)
und Zielerkennung mit Leitweglenkung aus der Postleitzahl würde
z. B. mit Hilfe der "message-switching-technique" möglich sein.
Der möglichen Vielfalt sind sicherlich noch lange keine Grenzen
gesetzt.

TELEFAX, GRUPPE 4
Betrieb nur in Digitalnetzen
Formularaufteilung?
Kopf*)
Schreib-
bereich
Individuell*)
Übertragung wie:
Telefax 4
Teletex
Telefax 4
*) Formulardruck
Verfahren setzt voraus:
- Einheitliche Proeduren für Telefax 4/Teletex
- "Multidienst"-Endeinrichtung

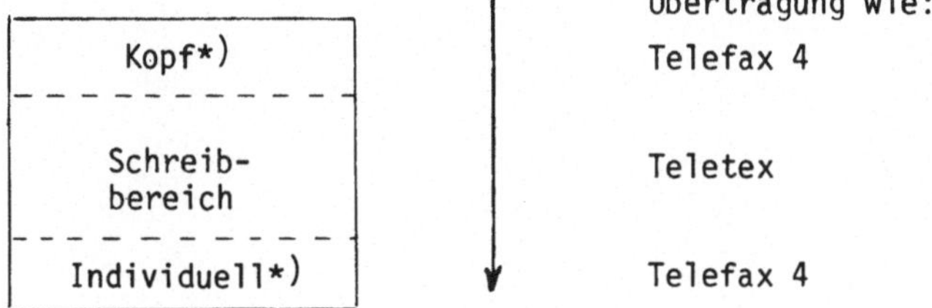

Bild 32

Telecommunications

Veröffentlichungen des/Publications of the
Münchner Kreis
Übernationale Vereinigung für
Kommunikationsforschung
Supranational Association for
Communications Research

Band /Volume 4
Telekommunikation für Bildung und Ausbildung
Telecommunication for Education and Vocational Training

Vorträge des vom 11.–12. Juni 1980 zur
VISODATA '80 in München abgehaltenen
Kongresses
Proceedings of a Congress Held in Munich
During VISODATA '80, June 11–12, 1980
Herausgeber/Editor: K. H. Vöge
1981. 13 Abbildungen. VII, 112 Seiten
(10 Seiten in Englisch)
DM 30,–
ISBN 3-540-10645-6

Telekommunikation für Bildung und Ausbildung wird schwerpunktmäßig dargestellt in zwei Übersichtsvorträgen, sechs Fallbeispielen und einer politischen Podiumsdiskussion. Ziel ist, den gegenwärtigen Zustand des Spannungsfelds zwischen Bildungsanspruch und Bildungsvermittlung mittels elektronischer Hilfsmittel zu beleuchten. Dabei soll deutlich werden, in wieweit Telekommunikation nach Meinung der Fachleute heute und in absehbarer Zukunft überhaupt in der Lage oder sogar zwingend notwendig ist, Bildung und Ausbildung zu vermitteln. Schließlich war die seit Jahren festgefahrene Diskussion zur Bildungstechnologie sowohl bei der Bildungsverwaltung und Bildungsausführung als auch bei ihr selbst neu zu beleben.

Band/Volume 3
Telekommunikation für den Menschen
Human Aspects of Telecommunication

Individuelle und gesellschaftliche Wirkungen
Individual and Social Consequences

Vorträge des Kongresses 29.–31. Oktober,
1979, München
Proceedings of the Congress October 29–31,
1979, Munich
Herausgeber/Editor: E. Witte
1980. 71 Abbildungen, 13 Tabellen.
XX, 335 Seiten (52 Seiten in Englisch)
DM 58,–
ISBN 3-540-10036-9

Mit diesem Kongress wurde versucht, eine umfassende Bestandsaufnahme der Telekommunikation im Hinblick auf die individuellen und gesellschaftlichen Wirkungen vorzunehmen, und zwar unter vielseitigen Aspekten, insbesondere aus der Sicht der Wissenschaften, der Medien, des Kommunikations- und Arbeitsmarktes sowie der Politik. Im ersten Teil des Kongresses standen die Anforderungen des Menschen gegenüber der technischen Ausgestaltung von Geräten und Prozeduren im Vordergrund. Im zweiten Teil wurden Probleme der individuellen Nutzung der Telekommunikation behandelt. Dabei ging es nicht lediglich um Marktanalysen und Akzeptanzuntersuchungen, sondern vor allem auch um die Frage, inwieweit die Bedürfnisse des einzelnen Menschen in neuen Kommunikationssystemen berücksichtigt werden können. Im letzten Teil wurden die gesellschaftlichen Wirkungen der Telekommunikation umfassend und kritisch diskutiert.

Springer-Verlag
Berlin
Heidelberg
New York